PRAISE FOR
## *From the Potato to Star Trek and Beyond*
**Book One of the**
*Treks Beyond the Great Potato* **Series**

**I wholeheartedly recommend this book with 5 full stars, shining brightly!**
Great read. Some thoughts and reminiscences from a well lived (and slightly worn) life!

I'm sincerely glad that I read this book! The first page left me with a lump in my throat. The final page left me (literally) with tears in my eyes...He is so compelling that by the time he is telling us about his father, mother, and grandfather's hernia, I'm committed and genuinely care about them, and I found myself rooting for their success even though they have already been gone by the time the book was written...He also describes the 'potato', but not until you're 66% through the book, and it is the absolute LAST thing that you could possibly imagine it to be...The book is genuine, it is honest, and it is him!...I will never watch 'The Tholian Web' again without thinking about an orbital rocket defense network, how many mirrors and how much gas it takes to make a laser, or the dangers of hidden rocks and waterfalls in a river again...And I appreciate that...

— Thaddeus Tuffentsamer, *Daily Star Trek News*. Amazon Vine Voice Review.

**In his captivating memoir,** *From The Potato to Star Trek and Beyond***, Chester L. Richards invites readers on a rollercoaster ride through a life that defies convention.**

Richards, a seasoned rocket scientist, shares a series of enthralling stories that not only detail his experiences with the unpredictable and the extraordinary but also serve as a testament to the resilience of the human spirit. The title of the book itself, which combines seemingly unrelated concepts like potatoes and Star Trek, alludes to the wide variety of experiences that take place inside its pages. Richards skillfully crafts a story that flows from almost fatal meetings with "The Potato" to unplanned contacts with Gene Roddenberry, the man responsible for creating Star Trek. The commonplace and the remarkable are juxtaposed to produce a lively and captivating story that has readers flipping pages with eagerness."

— Deb Patrick, *TheBookReview.com*

**Wisdom of Space and Time.**

Author Richards' memoir is classic storytelling through the eras. I liked the black and white photos of iconic moments coloured by the words and my imagination. I wanted to know the author beyond his external stories and he delivered his inner fears and triumphs quenching my curiosities about his life motives. I recommend this book.

— Tom Dutta, Host *The Quiet Warrior Show, TedTalks* speaker, Author, Business Consultant

**What a beautiful book!**

I returned to Solvang yesterday, following a three week absence, to find a copy of Chet's book. I could hardly wait to read it, and started flipping through some of the stories immediately. A modern epic!

— Jeffery J. Puschell, Ph.D. Chief Engineer, ElectroOptical/ InfraRed Payloads at Northrop Grumman Space Park, AIAA Fellow (American Institute of Aeronautics and Astronautics), SPIE Fellow (Society of Photo Instrumentation Engineers)

**A Riveting Journey of a Rocket Scientist's Memoirs**

Not only is *From The Potato to Star Trek and Beyond* a memoir, but it's also an ode to a life well lived and proof that any experience—no matter how difficult—can be seen as an adventure. Chester L. Richards encourages readers to consider their own travels, discovering delight in the act of storytelling and motivation in the face of hardship. For those who are in search of a story that defies convention and provides an insight into the remarkable life of a rocket scientist who dared to go where few others have ventured, this memoir is a must-read.

— Deb Patrick, *TheBookRevue.com*

**What can I tell you about my friend of decades, Chester L. Richards? He's a fine musician, a rocket scientist, a wonderful writer and an adventurer, but most of all, he married well.**

— Marilyn Scott-Waters, Writer and Illustrator Her works include: *The Toymaker: Paper Toys That You Can Make Yourself; Haunted Histories; The Search For Vile Things.*

**I loved** *From The Potato to Star Trek and Beyond, Memoirs of a Rocket Scientist* **by Chester L. Richards**

The author made a conscious decision early on to seek adventure in his life, and has he ever. Rather than having a beginning, middle, and an end, this book is a series of stand-alone essays about those adventures...He starts right off with a chapter from the title. When he was a graduate student, he and his friend Judy decided to write a script for Star Trek. Much to their surprise the powers-that-be liked it. He and Judy were swept up into the vortex that was TV production in the 1960s. The experience was enough to convince him that he didn't want to be a professional writer and to convince Judy that she did.

Adventures that were more physical involved paddling down unexplored rivers in Africa, another thing I love reading about (but would not want to do).

I particularly enjoyed the chapters that involved engineering. My husband was a structural engineer and he also had fascinating stories to tell, something I didn't expect from what I had thought of as a dry enterprise.

Richards' final story is called 'The Original War Baby' (World War II, for you youngsters). Unfortunately for him, I'm a week older than he is, so I'm taking that title away from him."

> — Caroline McCollough. From "Page Turners," July 2023 issue of *Mensa Bulletin*.

# The Trek Continues

## More Memoirs of a Rocket Scientist

Book 2 of the Treks Beyond the Great Potato Series

### Chester L. Richards

a Pawpress book

# The Trek Continues: More Memoirs of a Rocket Scientist
## Book 2 of the *Treks Beyond the Great Potato* Series
Chester L. Richards
Published by Pawpress
Edited by Ina Hillebrandt
Cover art by Eric Labacz
Interior Design by Corrin Hoppe, Pedernales Publishing, LLC
© 2025 by Chester L. Richards

All rights reserved. No part of this book may be reproduced or transmitted in this or any other universe by any means, electronic or mechanical, including photocopying, recording, or by any information storage and retrieval system, without written permission from the publisher, except for brief excerpts for reviews.

All photos either taken by the author, licensed or used with permission. For a complete list of photo credits, please see Photo Credits at the end of the book.

For information on reprints, bulk purchase, or available related performance and merchandising properties, and licensing questions, please contact the publisher via email: annap@InasPawprints.com. Or mail:
Pawpress
Brentwood Village .200 S. Barrington Ave. Ste. # 492213
Los Angeles, CA 90049
https://InasPawprints.com

Publisher's Cataloging-In-Publication Data

Names: Richards, Chester L., author.
Title: The trek continues : more memoirs of a rocket scientist / Chester L. Richards.
Description: Los Angeles, CA : Pawpress, [2025] | Series: Richards, Chester L. Treks beyond the Great Potato ; bk. 2. | "A Pawpress book."
Identifiers: LCCN: 2025903532 | ISBN: 9781880882344 (trade paperback) | 9781880882351 (hardcover) | 9781880882368 (ebook)
Subjects: LCSH: Richards, Chester L. | Authors, American--21st century--Biography. | Aerospace engineers--United States--Biography. | Adventure and adventurers--United States--Biography. | Extreme sports. | Rocketry. | Loss (Psychology) | Human-animal relationships. | LCGFT: Autobiographies. | BISAC: BIOGRAPHY & AUTOBIOGRAPHY / Personal Memoirs. | SPORTS & RECREATION / Water Sports / General. | SCIENCE / Space Science / General.
Classification: LCC: PS3618.I342 Z46 2025 | DDC: 813/.6--dc23

a Pawpress book
Printed in the U.S.A.

*To my parents, who showed me how to live*

# Contents

Prologue ..................................................................... 1
   Star Trek— The Trek Continues ........................ 1
Adventures Of A Water Baby ..............................11
   The Contract ................................................... 13
   Awash ............................................................. 19
   Cold War ........................................................ 32
   Double Trouble—The Omo ........................... 42
   Mythology ...................................................... 52
   Bugs ............................................................... 59
   The Fairy ........................................................ 66
   Sea Story ........................................................ 75
   The Perfect Wave ......................................... 105
Every Day Is An Adventure ................................ 117
   White Knuckle Flight ................................... 121
   My Hero ....................................................... 128
   Diversion ...................................................... 132
   The Bully ..................................................... 135
   Zot ............................................................... 137
   Failure .......................................................... 143
   Bug At The Beach ........................................ 176
   Thieves ......................................................... 194
   Poison .......................................................... 199
   The Woolsey Fire ......................................... 207

Friends and Fiends ............................................................. 215
   War Story ..................................................................... 217
   The Mossbauer Effect ................................................239
   Gold Fever ................................................................... 247
   Freeman .......................................................................256
   Life Savers .................................................................. 266
   The Rocket Scientist ..................................................275
Family ....................................................................................285
   Heritage ...................................................................... 287
   Some Assembly Required .......................................293
   Mom's Home Cooking ............................................. 301
Sarah .................................................................................... 309
   The Wedding ..............................................................311
   The Wedding, Part Two - Diplomacy ............... 315
   Hector ......................................................................... 319
   Kittens ..........................................................................328
   Moments With Sarah ................................................345
   Musical Gifts ............................................................. 346
   The Raid ...................................................................... 348
   The Horse ...................................................................350
   General Richards ....................................................... 352
   Scary ............................................................................ 354
   Wingnuts .....................................................................355
About the Author .............................................................359
Photo Credits .................................................................... 360

# The Trek Continues

WRITTEN BY
JUDY BURNS
AND
CHET RICHARDS

# PROLOGUE

## STAR TREK—THE TREK CONTINUES

INT. A spacious living room in a suburban Southern California home - NIGHT - Nov 15, 1968

WHILE MY FRIEND JUDY GREETS NEW GUESTS, early arrivals mill about her parents' living room, chatting over snacks. The mood is festive, the place packed. Friends and family have come to help us celebrate — Judy and I had come up with the original story and collaborated in writing *"The Tholian Web,"* an episode of the Original *Star Trek®* television series. Now we're about to see the show we'd created. For the first time. On live TV. With a roomful of expectant friends and family.

And I'm saying very little. To be honest, I can't wait to see how our words will turn out onscreen. We know the folks at the Star Trek offices liked the show. But this is different. This is the first product I'd had a role in putting out to the world. How would "real" people react?

As I'm waiting nervously for show time, a movement catches my eye — Judy's father, settling into one of the two chairs reserved for him and Judy's mom. A tall, rangy man, Mr. Burns had always commanded respect. Beginning as a police officer in Oklahoma, he'd moved on to a post that gave him a much broader mandate. In fact, he'd become an Oklahoma Ranger, teaming up with members of their counterparts the Texas Rangers on various occasions. In the thick of it for years, he had his share of tales of derring-do to tell. Among other things, he was, for a while, partnered with Frank Hamer, the Texas Ranger who took down Bonnie and Clyde. Retired now from on-the-job injuries, this evening must be as much a gratification for him as for Judy and myself. And of course for Judy's mom.

"Wait!" one of the guests asks. Racing out of the room, she returns quickly, bowl of popcorn topped up. Everyone else signals they're set.

It's time. Judy's mom and dad settled, the rest of us find a spot on the floor, snuggling up together to somehow fit into the space. Eyes look up, lock on the TV. The air in the room is electric. A friend turns on the television — a color TV, rare at the time. The screen comes alive. A cheer goes up — it's the opening tune and scene: the Star Ship Enterprise approaching against a backdrop of space and stars. Judy's name and mine appear on the screen. Wild applause!

CUT TO:

WE ARE ON THE BRIDGE of the Enterprise. The ship is tasked to find the missing Defiant, a sister ship. They do find it, but the Defiant keeps warping into and out of a parallel universe — seemingly a ghost ship. Captain Kirk leads a boarding party to the Defiant where they discover the crew is indeed dead.

*All this is familiar. After all, Judy and I wrote it! But it is fascinating to see how the actors and production people bring our words to life.*

## PROLOGUE

Captain Kirk is lost while being transported back to the Enterprise. The crew mourns at his memorial service. But then, he reappears — a spectral figure haunting the bridge and corridors! Is he somehow alive? Is he trapped in this strange region of space? Can he be recovered? Or is he really a ghost, truly dead and forever doomed to wander the ship, an apparition passing through the walls of the Enterprise. The crew of the Enterprise is terrified.

Meanwhile, the Enterprise faces a deadly serious problem: A Tholian ship arrives and a strange, crystal-like, being appears on the screen. He is anything but friendly.

It isn't long before everyone's cheering the good guys, booing the Tholians, and bombarding the screen with popcorn. People are making comments, laughing. In awe at the wonderful graphics. But oh, those clunky space suits...

The mood changes. The room becomes tense. Clearly, all hell is about to break loose. After the combat skirmish both the Enterprise and the Tholian ship are immobilized and must lick their wounds until repairs can be made. Then the Tholians deploy something new: a web of energy to envelop and capture the Enterprise...

*Over dinner at the Mongolian Barbeque Mike Minor had explained the exacting process of the web's creation. But now I finally get to see his wonderful stop-action animation of the web as it is being woven. I am really impressed.*

A few more scenes and the show is over. A cheer and applause charge the air. Our friends get to their feet and head to the kitchen for more refreshments. Judy and I join them, and are inundated with congratulations. Eventually people begin taking their leave.

The evening has been a triumph. I hadn't realized how wound up I was, wondering how it would go. Now, with everyone's enthusiastic reactions, I can relax and enjoy the satisfaction of a job well done.

THE HOUSE IS QUIET. Time to join Judy on the sofa for post-mortems. We laugh at some of the stuff that had been on the screen. There was that scene where Spock and McCoy are reviewing Kirk's video. "The dialog didn't make sense," I say. "How did that happen?" Judy fills me in. She had frequented the set often during *Star Trek's* shooting.

*I had only visited once during those times. That was not exactly a thrilling experience. Mostly people sat around for hours while the lighting technicians did their magic. Somehow the light aboard the USS Enterprise seemed to emanate uniformly from the walls themselves. You have to look very carefully at the final product to notice any shadows.*

About the dialog that didn't make sense: It turns out, Judy informs me, the actors didn't like their parts in the scene. So they swapped each other's dialog. Of course this completely wrecked the scene's logic. Judy caught the script trade at the time but nobody else at *Star Trek* apparently did. We have a good laugh.

I tell Judy that it was wonderful how the actors ran with the humor in a later scene. I know it had been a difficult one for her to write. Unhappy with her relatively flat first version, Judy tossed it and replaced it with a witty gem — a little masterpiece. The scene steals the show. Our friends bombarded the TV with a hail of popcorn when viewing it. They obviously agreed.

PROLOGUE

# FADE OUT

AND SO FOR A BRIEF WHILE I was a celebrity, if only among friends. But then, being in the midst of graduate school studies, it was time for me to go back to the routine business of learning.

For Judy, though, life changed. A few days after "The Tholian Web" episode aired I received a phone call from Bruce Geller, the producer of *Mission Impossible*. He offered me a job as a staff writer. I begged off. The real brains behind our success was Judy, I told him. I gave him her phone number. She went to work at *Mission*. This opportunity propelled my talented friend into a highly successful career as a television and movie screenwriter and story editor, and later a producer for television. The very popular *MacGyver* was just one of her successes.

Clearly, *Star Trek* turned out to be an amazing experience for me, too. For one thing, it proved profitable — paying for a couple of years of graduate school. Moreover, there were the residuals. Every year for half a century thereafter a check arrived from the Writers Guild.

Usually the residual checks Judy and I received were no great amount, though there were occasional pleasant surprises. My favorite of these came when the later *Star Trek* series used my idea of a personal force field for excursions outside the ship. Away with those tacky silver lame space suits!

I'm sure the actors in the later shows also appreciated the costume change. The *Web* actors had to be literally sewn into those suits, much like Elizabethan courtiers with their elaborate apparel. And for our actors, like for those earlier folks, nature calls were definitely a problem. Someone liked those space suits, though. Dr. McCoy's sold at Christie's for $144,000.

More fallout from the episode: As a precaution, Captain Kirk had left final instructions for Spock and McCoy. Their review of his intentions resolved the escalating conflict between them. Now they could once again work in harmony. This incident was later taught at the Law School of Texas Tech University as the first example of the use of a video recording for a last will. Cool!

Then, too, there was the Tholian Web itself. Our invention of this motivating device — to keep the characters and the audience in suspense — propelled the show forward. As I wrote in the first book of this series, *From The Potato To Star Trek and Beyond,* because of Mike Minor's extraordinary stop action animation to create the Web, the episode was nominated for an Emmy for special effects. It helped make his career as a premiere Hollywood art director.

And since my first book was published, I've again found, importantly among the personal residuals of our work, there are the fans, always the amazing fans. They are everywhere. Even my doctors are devoted fans. That was a surprise. It surprised my doctors, too, when they discovered my connection with *Star Trek.* Not considering myself a writer, I'd never really talked about my work on the *Star Trek* script before the book came out. But I was so pleased, and have remained grateful knowing that the episode made a hit with the Trekkie community.

As I mentioned in the first book, the fans nominated our episode for a Hugo Award, one of the great honors of science fiction. And *The Tholian Web* remains in the *Original Star Trek's* top ten, its various elements debated by fans to this day.

## PROLOGUE

And if you'll indulge me, I'm proud to again recount this moment. For several years business took me frequently to Washington. Weekends I visited the museums on the Mall. One day I climbed the steps of the American History Museum. The very first exhibit on the main floor was an exposition of Star Trek memorabilia. And there, prominently displayed right in the middle, was a copy of *The Tholian Web's* rainbow shooting script. We had made it, center stage, to the Smithsonian!

### The Legacy

AS SOME OF YOU KNOW, *Star Trek* had been rescued by the fans' enthusiasm after having been canceled at the end of season two. But just one more year, the Powers-That-Be at the Network said, then enough of this kids' stuff. Those potentates had no idea what was to come, no idea that *Star Trek* had already begun a cultural revolution. And that the iconic show would still flourish more than a half century later. Nor could any of us suspect the series would change the world.

I am pleased Judy Burns and I played our small part in this great endeavor. As a lark for me, but something much more for Judy, we had simply written a speculative script with the hope that the show would go back on the air for a third season.

The behind-the-scenes roller coaster of our involvement with the *Star Trek* show is told in considerable detail in my first book. And now, for those who have never seen it, the episode is easily accessible for viewing on the internet. Non-Trekkie note: Beware, if you decide to watch. It has some really scary stuff!

## THE TREK CONTINUES

AMONG MY MOST TREASURED "RESIDUALS": Judy gave me a valuable gift. I discovered I could write. Observing her at work was a powerful teacher. Judy has the gift. From her I absorbed many lessons as we wrestled with problems during the evolution of the story and script. I watched as Judy turned a routine encounter between characters into a work of art. But it took more to get me to put pen to paper.

First, Judy's screenwriting teacher, Bob Duncan, told me, seeing our script, "You should write." I realized to write, I had to have something to say. And so Star Trek really propelled me on a new path – I knew I had to DO things, have experiences, so that I would have something to write about. Thus it was that I embarked on putting adventures into my life.

During my wilderness excursion types of adventures, I kept a journal. Around the campfire in the evening I entertained my companions recounting the hazards of the day. They kept me honest, for there are no tougher critics than those who were there.

One day, during these times of exploration, eureka! The most thrilling adventure of all — I met and wed the remarkable Sarah. You might have read some of the experiences Sarah inspired me to write, finally, in my first book. And if so, dear reader, you will remember the moment the pen to paper process began. The moment nineteen years ago, when suddenly, after 20 years of sharing my life with this extraordinary lady, everything changed forever. I still feel the shock I felt then — my dear wife collapsed in front of me, her heart stopped.

But, you might also recall, losing my dear Sarah taught me a life-saving lesson. Writing the stories she loved helped lift me up from the despair, and heal the wound.

## PROLOGUE

Now the stories keep bubbling up, and I invite you to come with me on more hair-raising adventures, the kind that keep contriving to almost kill me. We'll meet more of the fascinating people I've been privileged to know, work with and count as lifelong friends. And, in this volume you'll find a series of encounters with fuzzy and warm, as well as frightening fellow critters on different continents — multi-legged, scaly, winged, and fabulous furry friends — who have graced my life.

This time I've written more about my lovely Sarah. She isn't gone. Sarah is present in each story — her spirit guiding me to find the words, as before, and more often throughout the book, you'll find her playing a role in the action.

You'll also see the section wrapping up this volume is devoted to stories and vignettes featuring Sarah as the leading lady — the opera soprano with rare gifts; stories showing her uncanny ability to talk with the animals; to spot — and foil — the bad guys; be a wonderful friend to lucky humans who came into her orbit; and about wingnuts.

I must say thank you again to the great lady, my dear Sarah, for being my muse and companion, for reminding me that everything in life is an adventure, to be recorded as it comes up in life or in my memory banks.

As with my first book, each story here is a letter of love to Sarah. Each is an adventure. Each stands complete by itself, meant to be read in any order you might choose. Each is meant to entertain or inform. Each is true.

*–Chester L. Richards*
*Oct. 2025, Thousand Oaks, CA*

*Whitewater rafting – Big Hole at Crystal, Colorado River, Grand Canyon*

# ADVENTURES OF A WATER BABY

IN CALIFORNIA AS A YOUNG BOY, I grew up with access to the beach and the delights of the surf, immediately at home in breaking waves and rough water. After all, I'd learned to swim at an early age. And lucky enough to be properly taught, I was never afraid to dive below the surface.

It was inevitable that I took up surfing, and later whitewater rafting and kayaking. For a decade I sought out the most challenging rivers.

Years after early surfing adventures, I most definitely felt the call to a whole new set of expeditions connected with water: sailing across deep waters with wind in the rigging, and blowing through the full head of hair I sported at the time.

Welcome aboard, dear reader! Let me steer you first to further adventures on, in and adjacent to the waters of a never before explored area of the world: the Omo River in Ethiopia. It all started with a fateful piece of paper.

*Mursi Tribesmen still Neolithic, Omo Expedition 1978*

# THE CONTRACT

Before my excursion down the Omo River in Ethiopia I had to sign a contract. It waived liability for various hazards. My friend Tasso glanced at the paperwork and warned me off this foolishness. When a revolution and civil war broke out in Ethiopia a few days before I was to leave he intensified his cautions.

Now Tasso had experience in these matters. Having survived a typical Balkan political fracas (a family affair — one of his uncles executed a coup-d'état against another of his uncles), and having had his destroyer shot out from under him in a naval engagement in a different war, he had developed some practical wisdom about that part of the world.

I did listen to Tasso, but by the time we talked I had too much invested to back out. I had paid my money, I had undergone rigorous physical therapy to repair an injury incurred the previous winter, I had suffered the miserable after-effects of multiple injections — cholera, yellow fever, smallpox, tetanus, diphtheria, polio and hepatitis — to protect myself against the most disease-ridden portion of the planet. But most of all, I had signed THE CONTRACT, and contracts are sacred with me.

This contract, with SOBEK EXPEDITIONS INC., did promise to provide local transportation, appropriate equipment and expert guidance for the expedition. But mostly its intent was to protect SOBEK itself from liability. As such it gave fair warning of the expected hazards. What hazards? Read on, for here is the core of the mandatory signed agreement:

---

*5. Participant acknowledges that he is aware:*

*(a) Of all the dangers and risks set forth below;*

*(b) That the Expedition by its nature may have other dangers and risks connected with it;*

*(c) That the boats in which the Expedition party will ride will pass through many extremes of water conditions, from calm and slow-moving water to very swift and turbulent water called "rapids" which pass over or around rocks and branches and waterfalls;*

*(d) That these conditions vary depending upon weather and water levels;*

*(e) That because of the many types of water through which the Expedition party will ride and considering the normal dangers of riding a boat, everyone riding in the river boats is subject to being jarred, bumped and splashed upon, and that the boat can lunge suddenly forward, up or down, or from side to side, or may strike some rocks or other obstruction and that a person could be thrown out of the boat and that the boat could turn over; and*

*(f) That the river passes through otherwise unaccessible terrain where dangerous animals, reptiles and insects may be found, where there are cliffs and hazardous terrain, and where there is the possibility of misunderstandings or hostilities with individuals from local tribes.*

## THE CONTRACT

*6. ASSUMPTION OF RISKS: Participant is aware and agrees to assume all the risks and dangers of this river trip, including but not limited to those set forth herein, and Participant agrees not to hold any of the other parties, their employees or agents, liable or responsible for any injury to person or property, or for any damage suffered as a result of such injury. There are four major types of RISKS:*

### A. RIVER RISKS.

*The first category is the river itself. At various points along the river, the Expedition will encounter rapids and "white waters", some of which will be extremely hazardous. At any dangerous points, SOBEK will give the Participant the option of either "shooting the rapids" or walking along the bank of the river and meeting the others at a point where the river becomes once again more easily navigable. At other points along the river, the boats will pass over or through small waterfalls. In addition, there may be rapid changes in the water level of the river due to the opening or closing of any dams or changes in weather conditions. There is also the danger of the boats striking submerged rocks, or hanging or fallen or falling trees, or other obstructions. Of course, there is always the danger of Participant being thrown from the boat.*

### B. WILDLIFE RISKS.

*The second category of risks includes the possibility of encountering wild and dangerous animals, reptiles and insects during the river trip. These may include crocodiles, hippopotami, poisonous snakes, buffalos, hyena, large cats, monkeys, hogs, lizards, and disease-carrying insects.*

## C. TRIBAL RISKS.

*The people who may be encountered along the river form the next category of risks. Individuals from local tribes and communities may try to harm those in the Expedition due to misunderstandings or unknown hostilities. Also harm may come from local bandits known as "Shifta."*

## D. LAND RISKS.

*The various land activities form the final category. While ashore during the Expedition, there will be places where those in the Expedition can hike over the terrain. The terrain will include cliffs, ravines, rough rocks and other dangerous areas. There is always the possibility of slipping on rocks or mud. There may also be opportunities of climbing over steep and mountainous terrain including cliffs and ledges. In hiking on adjacent plains, dangers from holes and brush also exist. Occasionally the river may become unpassable. At those times, everyone in the Expedition will be required to carry all the equipment and the boats to a point further down the river where the water is once again navigable.*

## E. MISCELLANEOUS RISKS.

*A few other miscellaneous risks include thunderstorms, extreme heat, government restraints or acts of other authorities, acts of God, and civil or foreign wars.*

---

Who with adventure in their souls could resist?

But SOBEK wasn't kidding. They were dead serious. I later discovered that on two previous SOBEK expeditions in the area there had been loss of life, to the river and to the Shifta.

# THE CONTRACT

*A Nile crocodile on the bank of the Omo. Adult Nile crocs like this are 10 to 15 feet long on average. This one was a rare 25 feet long. I shot this photo from our raft before he woke up and dashed for the water. The photo was taken with a telephoto lens in deep canyon gloom while we were riding a fast current.*

*Runup to Omo Expedition, Awash River*

# AWASH

As a warm-up for the Omo River expedition we set up our first camp at the Great Falls of the Awash River. The Awash threads through the Afar Triangle, a low-lying desert region extending from the Ethiopian highlands to Lake Abbe in Djibouti. This gave us a few days to build the team, explore the area, and drift for two interesting days down the river.

The Awash has little to offer the Danakil people there except water, water to nourish cattle and the homicidal young men who take "testicular trophies" from the young men of rival tribes. At least that's the way it used to be when I was there, half a century ago.

The first day of our excursion was l-o-n-g. In the morning we left Addis Ababa high on the central massif of Ethiopia. It was good to leave that region behind for it was now immersed in the first days of a bloody coup. In the end, multitudes would be murdered by the new government. Many years later a witness, who was then a child, told me about the horror. In accordance with Stalin's quip that "A single death is a tragedy. A million deaths is a statistic," official thugs carried out the Junta's orders. On frequent occasions the silence of the night would be shattered by gunshots. Soldiers took groups of people, never just one, from their homes and executed them, leaving the bodies heaped in a nearby field. Over the following years the death toll would be in the millions.

While driving away from Addis along the highway, we had a nightmare encounter with soldiers of the new regime -- the moment *when readers of my first book might recall one of these armed gunmen spotted a camera hanging down over my stomach from a strap around my neck. Glaring, he pointed his rifle at my belly. It took two of us gesturing frantically, trying to convince him without words we were just there for adventure. A very hairy moment, that. For now, thank heaven, that was all behind us.*

## From My Journal
*Written as the events described unfolded*
*30 September, 1976*

WE STARTED THE LONG DESCENT DOWN from the massif. Midway down the slope of the steep escarpment we saw below us a small, walled city, which we approached along the switch-backed road. This was the medieval town of Nazret. Lush Bougainvillea completely engulfed the high walls around the old settlement. As we passed through the old part of town we were surrounded by houses festooned with these brilliant flowers and vines, radiant in the intense sunlight, with stupendous masses of glowing purple blossoms.

Leaving Nazret after being happily immersed in its glorious flower blankets, we arrived at the Awash River site late and immediately set up camp, after chasing away a band of hyenas. Our campsite was just above Awash Falls. The river above the falls is shallow and sufficiently wide that the falls stretch across the head of the canyon below for more than a hundred yards.

*The area is a National Park. In those days it was completely undeveloped. Only one small structure was there and a dirt road leading a short distance beside the river. These days the falls area is immaculately landscaped around a luxury hotel.*

Tired as we were after the day's drive, we enjoyed the evening, after chasing hyenas away from the camp.

Before retiring I walked down to the river and found a place to sit and simply enjoy the soft, mellow evening. Below the falls, I encountered fireflies for the first time since I was a boy. The tiny flickering lights darted in and out among the vegetation which grows up the low cliffs along the river. My long ago experience with these little creatures was on a drive through Midwestern summer woods. Some of those vast numbers of night jewels collided with our windshield, clouding it with soft light. We stopped to clear off the green glowing substance on the windshield so that Dad could once again see to drive.

The alien night sky, with the moon upside down in the north and strange southern constellations, catches our attention and is a topic of spirited conversation around the campfire. Night sounds are new, as well. Listen: The crickets here chirp with a higher, more mellow, pitch — quite different from home.

## 1 October

**The water in the Awash River** was down yesterday, and still down, so we decided on an alternative: a hike up to the rim of Fantale Crater. The crater is a caldera about two miles in diameter. It is located in the northern section of Awash Park. To get to the crater, we drove along a road that was at first paved, then became a dirt road of very primitive character. But what a treat! A number of different animals — wart hogs, ostriches, herds of kudu, oryx and other types of antelope appeared in clusters along the way.

## THE TREK CONTINUES

We hiked up the slope of the caldera about three miles, and climbed fifteen hundred feet. The climb was somewhat exhausting because of the broken trail we followed and also because of the very fast pace our leaders set.

Coming over the rim I was impressed with the lush, irregular landscape in the sunken land far below. Initially a deep overcast drowned the crater's floor in gloom, full of dark, richly colored shadows. After a while the sun came out and the caldera blossomed into an iridescent green and gold shimmer.

Totally frazzled after hiking back to the bus in the blistering equatorial sun, we were well ready for a five minute stop at Metahara, a whistlestop nearby, to tank up on Cokes. Driving on we passed a small troop of hyenas on the side road leading down to the campsite, perhaps the same troop that had visited us the previous night.

*Monkey on monkey time among Vervets*

# AWASH

We arrived back at camp in twilight. There we were greeted by visitors — unwelcome visitors. In fact, the camp was infested — with the band of Vervet Monkeys I mentioned to dear readers in my first book. Let me elaborate a bit here.

They were inside our tents. They were on top of our tents, collapsing one while we watched. They were rummaging in our kitchen box, flinging pots and pans around with a great clatter — which drove them into paroxysms of maniacal laughter. One of them grabbed an aluminum plate and tossed it like a Frisbee. It sailed past my head. Others, in gleeful imitation, also fished out plates and started flinging them, luckily not at one of us, but far out into the countryside. We counter-attacked, flapping our arms wildly and laughing maniacally ourselves, and soon sent the bandit marauders fleeing into the bush.

Cleanup took some time — us weary travelers weren't quite up for this. However...

*Who could resist?*

## 2 October

I AWOKE AT DAWN, dressed and wandered away from the camp to savor a bit of the dawn wilderness. Breakfast was still an hour ahead.

*Vultures on the sleeping tree*

Nearby I discovered a sleeping tree. Now dead, the tree had been stripped of all leaves, twigs and bark. A number of great black vultures perched in its many branches. These are giant birds. With wingspans 5- 9 feet, the larger birds match the size of our rare California Condors. We had seen them soaring high above us the day before, circling in thermals. The trees in the area are mostly clustered along the river where ground water is plentiful, so it is natural that somewhere around here would be the dormitory of the great birds.

# AWASH

Later in the morning I again hiked down the road, this time to the small exhibit center near the park entrance.

*Along the way I passed a tent site that was being set up awaiting that day's arrival of a Lindblad Tour. Each spacious tent had a soft mattress cot with fresh linens. Beside each cot was a side table. On the table was a small crystal vase with a fresh cut red rose.*

*So this is how the other half lives! I was amused for I was fully aware that sleeping on the unyielding earth, and enduring daily hardships, we were having a much better time.*

The untended center was a small, single room, concrete block building. Inside was a chart of rainfall in the area. The terrible drought that the country had endured in previous years was clearly plotted. At its worst, the rainfall total had fallen to only a couple of inches a year. Perhaps this drought was one of the causes of the political turmoil that Ethiopia was now suffering.

Outside the structure was a cage with wide spaced bars. The cage was inhabited by a leopard. Laying there he looked up at me with a sad expression. I stooped down and started talking to him. He purred in response. Even though his face was only a foot from mine I did not try to pet him – although it was a great temptation. Our long chat seemed to cheer him up, for he soon sat up and started nuzzling the bars of the cage, his purring growing louder. He was just a big kitty. There was intelligence in his eyes. I hope he finds a home very much better than where he is.

I returned to camp, joined everyone for breakfast. At last we were ready for the river. We started just below Awash Falls, a most remarkable place to put in. The falls stretch straight across the river, somewhat like a miniature version of Niagara

Falls in its proportions. The height of the falls might be thirty or forty feet, maybe somewhat more.

Above the falls the river drifts lazily across a flat prairie, with trees clustered along its banks. We prepped our three rafts shaded from the African sun by an ancient tree. Inflating and loading the rafts is never an easy task so we blessed the cooling mist of the falls and the darkness under the tree. After more than an hour of hard labor we were ready for the put-in. Off we went drifting downstream for a two day adventure.

Below the falls the river drops into a shallow canyon, which progressively deepens downstream until its depth becomes quite substantial. There are occasional open areas where the walls of the canyon recede and there is ready access from the terrain above. In these areas natives come down to the river for various activities.

As we drifted out from the narrow canyon, and around a bend, the riverbanks widened out into a gravel bar forming a gentle rapid. Although I could not see the cause of the gravel bar, it is most probable that a stream, flowing down from the hills to the right of the river, had, in a flash flood, created this miniature rapid.

*This is where we encountered the amazing Danikil women I'd told readers about in the first book of my memoirs singing in three-part harmony and intricate counterpoint the likes of which I'd never heard. And where we encountered another brush with death: three men, one of whom leveled his rifle at me when he saw my camera. My seatmate reached over and covered up the camera with her hat. Many salaams later, he lowered the rifle. I breathed a sigh of relief – a big sigh.*

All day at occasional intervals we could hear the cries of children, high up along the cliffs, commenting on the novelty of three boats filled with white men, drifting on the river.

Late in the afternoon we heard lots of chatter coming from downstream and again around a bend. We were the last boat and it turned out that the lead boat was passing a large group of boys who were bathing in the river.

The chatter changed into singing and waves of greeting. Much passing back and forth of Salaams. As we brought up the rear the boys became positively aggressive, chasing us along the bank and finally diving into the river and swimming after us. Unsure what they were up to, we started rowing hard and left them behind.

Our camp was set up near a hot spring, its Jacuzzi-sized pools overflowing down to the river with water at just the right temperature to ease aches and pains. We bathed for a long, luxurious time in the moonlight. Jim Slade, our leader, emphasized that this was a sacred spot. He pointed out the little twig figurines deposited as votive offerings along the travertine cliff wall behind the springs. I had seen the like before, at the juncture of the Colorado River and the Little Colorado in the Grand Canyon — a very sacred spot for the Indians, the place where Humans came to the Earth from the Other World. We were careful to leave a clean camp with the sand carefully brushed behind us.

## Sunday, 3 October

In the morning, as we worked our way down the short trail from the campsite to the river, our path was temporarily blocked by a huge spider web, occupied in its center by a giant orb spider. And I do, very much, mean giant. The spider was about the size of a small dinner plate. With its central body significantly larger than my fist, it was really more crab than spider. This spider was of the variety that feasts on small birds.

The immense, spring-steel-stranded web was slung between two trees on either side of the trail, and so blocked our passage. I must say that the web was beautiful beyond easy description. The low, early morning sun diffracted from its thick strands with dazzling, shimmering rainbows — the brilliant colors flashing back and forth along the strands with each minute disturbance in the air. One of the boatmen pulled out his knife and sawed away a portion of the web at its bottom so that we could duck under. The spider did not seem very perturbed. I'm sure that large animals, passing along the trail, cause about the same amount of damage. In any case, this most exotic of creatures would be able to easily repair the elaborate structure that had been constructed in just a few hours originally.

We kept a wary eye on the exotic arachnid as we proceeded. Thank heaven it did not jump out at any of us in rage.

BEFORE TAKEOUT, we had a pleasant few hours drifting on the river. Pleasant doesn't mean uneventful.

*There are the rapids. Most rapids form when debris washes down from a side ravine to dam up the river and form a long deep lake. Downstream of the dam the water in the rapid shallows. Here is the danger. The Awash is not deep. But its rapids are typically "boulder gardens." No big waves, but plenty tricky nonetheless. We descend into a kind of pinball machine, bouncing us from boulder to boulder.*

Because of the low water level we had some trouble negotiating a few of the rapids. In two rapids our boat got thoroughly wrapped on the rocks. Each time a delicate but slightly frantic rebalancing of the boat so as to work it free followed. Thank heaven no other mishaps occurred.

The pool upstream of one rapid teemed with fish. Here a multitude of kites milled around just above the water in a kind of animated lattice. Dazzling white feathered, their wings fringed with black, the birds were predatory angels. Each took its turn to skim the water and fly off gripping a still wriggling victim.

At one point the river headed straight for a river-carved cliff and then veered around a blind bend to the right. The gray-walled cliff was perhaps a hundred feet high before it merged back into the hill behind. Towards the top of the cliff was a ledge backed with shallow alcoves. As the river current drifted us towards the cliff a scout baboon, on the high ledge, caught sight of us and howled an alarm.

Within seconds a whole troop — dozens of baboons — appeared at the edge of the ledge. Hooting and hollering, they jumped up and down in an increasingly frantic tarantella as the current swept us ever closer. How these creatures reached this apparently inaccessible niche in the cliff remains a mystery.

Just as the river started to swing us to the right we came within range of a rain of projectiles. The baboons were hurling these strange dark missiles down at us — thereby giving the lie to those anthropologists who still maintained at the time that throwing weapons is a uniquely human trait.

Suddenly, someone called out that the missiles were baboon turds, obviously stored in great piles in their nesting perch. Baboons are evidently not very sanitary animals. Fortunately, they aren't good shots either. For all their efforts none of the nasty objects landed in our boats. We drifted on, hastened somewhat by the strong oar strokes of the boatmen.

At takeout, near a bridge, we were greeted by several Danakil watering their large herd of cattle at the river. When we came ashore the tribesmen hustled the cattle up the steep road to the highway and then returned to sit and watch the action. Takedown of the inflated boats required considerable effort. The boats had to be stripped of their gear, propped up on oars sticking out of the beach sand, washed thoroughly inside and out to remove the abrasive sand caked along the sides and bottom, and finally left to dry, deflated, rolled and tied up. For our three boats the whole process took substantially more than an hour.

Our chartered bus, which had been driven down to the river, was loaded and sent on its way. We had been instructed to walk up the road to the highway ahead of the bus to minimize its load — it was having enough trouble hauling itself up the steep incline with all our gear on board. As we reached the top of the rise we encountered a rather ancient-looking shepherdess hustling a herd of goats, together with a few sheep, across our path. Loose-jointed and flat-footed, she was probably the beneficiary of tens of thousands of miles of migration over the millennia of her predecessors in this part of the world. Clarice smiled at her. She smiled and came over, and after many Salaams and friendly smiles, she and Clarice exchanged kisses.

# AWASH

*Gelada Baboon Awash, Ethiopia, roaring his alarm Not our guy – I had no time to set up a photo. But this is a closeup of just what we were witnessing from the river below.*

# COLD WAR

**3 October. Sunday evening.**

At last we were ready to head out for the next leg of our adventure. A couple of hours after we'd joined our luggage, our bus carried us off the paved highway and proceeded for the rest of the afternoon on a seldom used trail. Our long trek led us into the wilderness. Down we descended through lush grasslands until we reached the desert floor of the Afar Triangle. Continuing to drive further into this desolate land, we were stopped in our tracks about fifteen miles northeast of Fantale Crater.

There, in the middle of the desert, we'd come upon a scene straight out of Hollywood — a huge, lush desert oasis complete with hot springs. Palms lined deep, sparkling, azure pools of water, some of them fed by low waterfalls and some by clear ground water.

One could almost visualize Dorothy Lamour spirited to this remote oasis by a handsome desert sheik on a fiery steed. We had our lovely expedition ladies, but the closest we had to a desert sheik was our bus driver. Somehow he was not quite from Central Casting.

Testing each of the pools, we discovered one had an ideal water temperature — that of a warm bath, or spa. Everyone rushed to make camp, stripped quickly, slid into this best of all possible pools, and proceeded to bathe the grime of our three days on the river out of our bodies.

*Lush palm trees in the middle of the desert hot springs oasis near Fantale Crater in Ethiopia.*

The awesome sunset — broken clouds ablaze in gold and scarlet — seemed enormous. As we gazed upwards, miles long flocks of swifts rocketed towards us from the northwest. They were nearing the end of their migration back to Africa for the winter. Whizzing by only a few feet over our heads, they were close enough that we could feel the faint buzz from their wings tickling our faces. Swifts are the fastest birds in the world and, in migration, they fly across the continents and the seas, day and night, without landing. How they do this without seeming to refuel and without rest is a puzzle.

Refreshed after bathing we gathered for dinner. A casserole had been prepared back at Addis then deep frozen and tucked into an ice chest. We retrieved it from the bus, gathered wood, and soon a blaze lit the growing dark. By now only a narrow band of red remained above the western horizon.

It took only a few minutes for the fire to warm the meal inside an iron kettle. Soon enough the savory aroma told us it was ready. We grabbed tin plates, dug in and again took our places around the fire.

Sitting on a log to my left was our bus driver. He was of indeterminate middle age, wilderness worn, and well experienced with local hazards. I was absorbed with the delicious meal and not paying attention to anything else. Suddenly the driver put his plate aside, jumped up, grabbed a flaming branch out of the fire, and leapt in front of me. Whack! Our driver brought his club down on a large, very nasty looking, very shiny black scorpion strutting along, tail arced up in attack position, only a few inches away from my bare toes. Then he pushed the still writhing thing into the flames. "Very bad!" he exclaimed. "Very deadly." Satisfied, he sat back down and continued his meal as if nothing much had happened.

Later, after supper, we again splashed into our now moonlit spa, delighting for another moment in this tropical paradise. Each took his turn occupying various parts of the inflowing waterfall, enjoying nature's massage. *Ahhhhhhh.*

Though the night for me began pleasantly enough, it ended with intense suffering. I had apparently developed a serious enteritis. Barely able to crawl out of my tent, I laid the rest of the night in the warm sand, unable to move, fluids exploding from every orifice. Not a pretty sight.

Getting ready in the wee hours before our departure back to Addis Ababa the next day, my companions discovered my plight, packed my gear, cleaned me up, dressed me, nursed me with life-giving broth, and helped me to the bus for the long ride home. "Chet's Revenge" my malady was called by those who later came down with a milder version.

We bussed out of the campsite at sunrise to see as much of the park as possible in prime viewing time. Our reward? Having gathered together for the night, the early morning herds of wildlife were much larger and not quite as shy, and there were even more species to see, though one we had encountered was missing. *Lest you wonder which, let me just say that no charming munitions were lobbed at us this day.*

We also saw herds of zebra, and several herds of possibly the most fascinating creatures of the trip so far — what I learned were stotting gazelles. A portion of one herd was running figure eights just as if they were on a closed racetrack. Their mode of locomotion is rather strange though undoubtedly effective. They don't run, they spring. It's as if they have steel coils attached to each foot. They boing up into the air like movie cartoon characters. Visualize a four-footed kangaroo. To start their movement they leap straight up, all four feet simultaneously off the ground. Then, when they touch down, they leap again with their hind legs phased slightly ahead of their forelegs. This propels them forward several feet at a bounce. The result is an extremely fast forward motion together with great maneuverability.

We also saw giant desert tortoises, at close range. In fact, we spotted a number of these impossibly huge creatures in a cluster close by. The largest had a shell whose top was well above the waist of one of our party who had gone over to get a close look. This monster must have been centuries old. Then there were the wart hogs. No one attempted close scrutiny of these extremely dangerous animals. We saw a family of three, mother and child huddled behind daddy for protection. Dad just glared at us, his haunches up, muscles tensed, ready to charge. We mere humans stayed safely on the bus when we got near these fearsome critters.

*Family of warthogs*

Despite this welcome interlude on our drive back to Addis, most of the trip was long and tedious for everyone. Especially me.

Finally, we came to the hotel in Addis! Even in my condition I joined the beeline to her door. Because of my illness I was moved from my old double into a single room on a different floor from the rest of the expedition, so that I could get the rest needed to recover. Dinner was out of the question. I slept.

When I awakened, it was dark outside. By day Addis Ababa belongs to its million human inhabitants. By night Addis is a city of a million feral dogs. Tribal packs collide in the streets, growling and snarling, then rebounding to again patrol their time-honored territories. Great waves of barking, yowling, yelping and howling erupt across the city, now from the east, then swinging around to come from the north. The waves delineate the geography of the city in phantom sonic shapes. They flow until they shatter on some geographical impediment, a hill or a structure. New waves spring up from the shards of the old and take on lives of their own as they begin their migrations across this city of creatures of the night. Humans cower safely behind their shielding walls. Despite the rare and fascinating cacophony, again I slept.

## Monday, October 4

The next morning, I had recovered enough to go down and get some breakfast. Meals at the hotel were only offered at specified times. Miss the time and go hungry, so everyone in the hotel migrated towards the dining room together. Walking some distance ahead of me down the corridor was a group of Chinese men who apparently were staying nearby. They reached the floor's lobby just after a similar group of Russians came in from the opposite wing. The Russians got in the elevator. Rather than join them, the Chinese elected to descend the long flights of stairs. *Hmmmm.*

A few minutes after I returned to my room I heard the delicious sounds of the troupe's master musicians tuning up for their performance that evening. The musical craftsmanship was extraordinary. Naturally I left my door open so as to gather in as much of the sweet sound as possible. The room attendant, seeing my open door, wandered in and started practicing his English. After a few pleasantries and a bit of cleanup he left and started to close the door. I signaled him to leave it open. He ambled off shaking his head about the preferences of this crazy American.

After a while curiosity overcame my still lingering weakness. I had to see where these tantalizing harmonies were coming from. A short way down the corridor a door was open for ventilation. Inside a Chinese musician was practicing. He was playing a familiar Bach passage, taken from a violin partita. This was on a traditional Chinese percussive stringed instrument. The sound was strange, but very sweet. As expected, he had total mastery of the instrument, so that the arpeggios just rippled out in waves. The tonality was something like that of a clavichord or spinet. I learned later that these entrancing musicians were The musicians of the Red Chinese Ballet troupe, here to perform live.

I had learned at breakfast that the Moscow Symphony Orchestra was also in town. I knew now that they were in the other wing, so I wandered down in that direction. I came to a room with its door also cracked open. By a wonderful coincidence, inside a master violinist was practicing the same Bach partita. I was fascinated to hear the same melody emanating simultaneously from very different cultures— different sounds, same tune.

Not yet recovered, the long walk was becoming too much. I drifted back to my room. Overtaken by fatigue, I lay back in my bed, falling asleep as I savored the sounds of the nearby Chinese artists. After a most refreshing nap, I wandered down to the lounge. The nearby bar was empty except for a few highly manicured women who waited listlessly, hoping for assignations to come.

I decided to take a seat in the hotel lounge — it was quiet, unlike the bar, with its loudspeaker incessantly blaring propaganda from the new revolutionary communist government. That afternoon proved fascinating. I found myself in an hours-long conversation with a gentleman who was the director of our medical advisory mission in India. As I recall, the project was under the aegis of the U.S. National Science Foundation. He talked about the view from a tall building in Benares, in what had become, in 1950, India's fourth largest state, Utter Pradesh. From the top of one of the tall buildings you could see a considerable stretch both ways along the Ganges River. What struck him were the people — vast multitudes of people. On some days more than a million human beings could be seen at a glance.

His job was supposed to be science. But mostly, he told me, it was dealing with dignitaries. It was always the same. A delegation of U.S. congressmen or agency heads would arrive. As they got off the plane he warned them not to drink the water. Naturally, the very first thing they did was head to the airport bar for refreshments. And they invariably drank the water. That's okay. The director had already built a week's recovery period into the schedule. We laughed at my all too fresh experience with Chet's Revenge, even though I had avoided at least knowingly ingesting the water.

IN THE EVENING WE SKIPPED the hotel's scheduled dinner and made an excursion to eat at Omar Khayyam's, reputed to be one of the best restaurants in town, to celebrate tomorrow's start of the real expedition. As it turned out Omar's was closed, so we went across the street to a pizzeria. Because of the size of our crowd we were ushered down into the basement where a large table awaited us. Twelve of us were seated at this Last Supper of our city sojourn.

## THE TREK CONTINUES

Our waiter handed us menus. While these had English translations, the waiter spoke only his native language, so John Yost, fluent in the Ethiopian language Amharic did the translations for the rest of us.

Kathy, a young biology student, started the ordering with her first selection, and John translated. Before Kathy could give the rest of the order to John, the waiter departed for the kitchen. He returned a few minutes later with a single plate of liver. "That's not what I ordered, I asked for noodles!" The waiter looked confused. Before he could get away he was bombarded with orders yelled out by everyone at the same time, in English: three chickens, two pizzas, veal, lasagna, spaghetti. It was too much. He brought a few dishes and let us choose. "Hey, what about my soup? I asked for the Minestrone soup." "You forgot the vegetables." "Four bottles of Sprite, please." John yelled for everyone to shut up. Waving his hand over the menu, he told the waiter to bring everything on it. The waiter nodded, making several trips to load up the center of the table. Everyone immediately grabbed what they wanted.

In the end, the food was quite tasty and we all got through our meal well fed. After finishing, John added up the order for the waiter.

Exhausted, I was glad to get back to the hotel, and to bed, once again to be serenaded by the wandering packs of feral dogs. Tomorrow we would head south to meet the Omo River.

### TUESDAY MORNING, OCTOBER 5

AS I WAS HEADING DOWN to join our group for our final breakfast before heading out to the Omo River, a strange thing happened. Unlike the day before, both the Chinese and the Russians arrived *simultaneously* at the elevator. They stood there glaring at each other, neither making a move. I ambled up and walked into the elevator, whereupon both groups crowded in with me. I was an American, a neutral party. And it appeared I had negotiated a temporary truce in a different kind of Cold War.

# DOUBLE TROUBLE—THE OMO

*Come with me again on the next adventure in our expedition in Ethiopia.*

Journal entry: Friday, October 8, 1976. Evening. Monsoon season.

The rains have swollen the great Omo, a river comparable to the Colorado, in a canyon even deeper than the Grand. Unexpected things happen on wild rivers.

### How to Name a Rapid

Today, shortly after starting, we entered a gorge. I hitched a ride in Jim Slade's boat, which was to lead. "Today is not primarily a whitewater trip," Slade told us at the outset. He did mention we would experience a few moderate rapids after lunch.

Slade and the other boatmen were not the least concerned about this morning's stretch of the river. Even though the water level was higher than they had previously experienced, the boatmen all expected that the river, for several miles to come, was going to be relatively mild. They did not even bother to lock down the lids on our "lunch boxes," large wooden affairs which stretch across the width of the rafts and hold our food and kitchen supplies.

## DOUBLE TROUBLE—THE OMO

A half hour or so into the gorge we came upon a series of modest rapids. We took some water aboard but had no significant trouble. Suddenly the river veered sharp right, then left, sluicing off the wall in a counterclockwise direction and leading down towards a substantial rapid. The sharp turns in the canyon hid the upcoming rapid. Its magnitude came as a shock. Which goes to show that a truly wild river, such as the Omo, always has something to teach the unwary.

As usual, our boats were strung out in two pairs of two. Each boat of a pair followed close enough behind to assist its companion in case of trouble. The second pair of boats stayed well back from the first so that they could learn from the experience of the first pair in case the lead boats got in trouble. This positioned the people in the second pair of boats not only to take a different route if necessary, but also to come to the rescue in case of lead boat accidents.

From the top of the rapid we could see several good size holes but there was no time to stop and investigate — we were already sliding down the tongue. Punching through the first big hole without much trouble, we easily skirted a second. Then, *bam!* Right down into a deep, dark keeper. The boat rocked up on the haystack, sticking there for a few moments, about to pull free. Suddenly, an extra surge of water picked us up and flung the boat topsy-turvy. We were swimming for our lives. Thank God for life preservers!

I surfaced under the overturned boat, in its air pocket, but easily worked my way clear. Clarice was not so lucky. She became trapped under the boat and remained there for some time, entangled in the dangling ropes. When she managed to emerge, our fellow passenger was badly shaken. Lou Steiger, an apprentice boatman, was the other passenger. He was aboard to learn the river from the master. Fortunately, he had no trouble getting clear.

Shortly after coming to the surface I turned around and watched the second boat in line drop into the same hole and flip. The two following boats learned from our example and skirted the hole.

The fast current quickly took us downstream. I worked my way upstream of the boat, this being the safest location. Debris began popping up — plastic containers filled with various foods and condiments materializing from the opaque, sediment-laden water. A wrapped cube of butter floated to the surface and submerged again, pulled down by the river's turbulence. Food containers and other items were popping up and going down again several yards off to one side of the overturned raft. I swam away from the raft and started corralling as many of these as I could. Within a minute or two, I had a growing collection of salvaged items, all floating in a nice cluster encircled by my right arm.

While engaged in my salvage task, I saw something out of the corner of my eye surface and quickly drop back out of sight. In an attempt to acquire the item, now hidden below the surface, I reached down with my left hand into the coffee-with-cream colored water. Lo and behold, the object was not manufactured, but alive. There are *things* in this river!

This thing had teeth, and lots of them. Very pointy and sharp, too.

So here's the situation: I'm floating downriver bouncing up and down in the tail waves of a big rapid. Water-polished rock cliffs rise straight out of the water on either side of the river, so there is no place to go but downstream. With my hand — and arm — part way up to my elbow, inside the mouth of some big creature of the river. And I've just realized that the creature must be a crocodile.

The boat flip had already flooded my body with adrenalin. Now I felt an extra burst of the stuff — a major hot flash. Everything slowed down, waaaayy dooowwwnn. Legend has it that in such a situation your life passes in front of you. Something like that happened to me, but I wasn't quite at Death's door — not yet. Instead, I found that I had plenty of time to analyze the situation. The memories that I needed at that desperate moment magically surfaced, complete with detailed images.

Part of my mind was sorting out the behavior of crocodiles. I understood that crocs clamp onto their victims and drag them down to the depths to drown. Then they can eat at their leisure. Accordingly, a crocodile has powerful muscles to keep its mouth clamped shut, and relatively weak ones to open it up. Any false move on my part, such as trying to pull my hand out, would surely trigger that bear trap mouth. What to do?

A part of me went back to graduate school. While I was enrolled in physics and engineering I also informally studied the new discipline of psychobiology, attending seminars and colloquia, and ransacking the library. I learned about reflexes and how they work. In reviewing that material in my mind's eye, I flashed on a picture in one of the books —a baby hanging, with both hands, suspended from a clothesline. The text described the surprising strength of the baby's grasping reflex. It also talked about a counter reflex. These reflexes are reciprocal and depend on the strength of the stimulus. Lightly press the center of the baby's hand and the baby will tightly grab your finger. Press hard and the baby's hand will fly away from your touch. The picture, the text and the context were all there, available to assist me.

Suddenly the answer dawned on me: most reflexes must behave much the same way as the baby's grasping reflex. Therefore, if the croc's normal jaw reflex is to clamp down with a mild inside stimulus, then its mouth must fly open when given a sufficiently unpleasant stimulus inside.

It was time to act! I balled up my hand and rammed my fist down the croc's gullet with all my strength. My arm was now deep inside the trap, the croc's teeth pressing against the back of my elbow. As I hoped, its mouth flew open with a gag reflex. I jerked my hand out of danger, scraping my palm along a row of teeth in the process. From start to the instant I punched the croc perhaps only one or two seconds had passed — a very long two seconds. I had been suspended outside of time. Now I was back. The river was flowing again.

In a Newtonian reaction to the force of my blow, the croc's head reared up out of the water. For a fleeting instant I got a clear view of its long, narrow, triangular skull before it dove back under water on its panicked flight out of the area, leaving behind only a wake. I, too, was not going to loiter there, so I abandoned my salvage and swam back to the overturned raft.

"There're crocodiles in this river!" I yelled out as I hurriedly made my way back to the relative safety of the boat. Even a modest sized crocodile is nothing to be trifled with. My companions, floating on the other side of the raft thought I was nuts. Crocs just aren't found in rapids. That is, they thought I was nuts until I raised my hand, bloody, with folds of flesh dangling down from four deep, and long, parallel slashes. Then all clustered tightly around the raft.

## DOUBLE TROUBLE—THE OMO

We landed downstream and righted the boats. I tried to help with this, but my last reserves of adrenalin suddenly ran out. Flopping down onto the sand, I laid there trembling, a mere rag doll.

After the boat repair we took stock. Our major losses were to our food supply. At least two large lunch boxes of food — about three weeks' worth — had become a sacrifice to the river god. Fortunately, we did not lose any kitchen utensils; pots and pans and the like are really important. All my personal gear came through intact and dry. Our only injuries were to our pride and our stomachs.

I suggested a name for this rapid: *Double Trouble.* The name stuck. So now I am a member of the select club of those who have named a geographic feature, however minor.

Later we stopped and had lunch at Cloudburst Falls. This gave us a chance to let our adrenalin replenish and to swap stories.

### How to Name a Hole

AFTER LUNCH WE DRIFTED a couple of minutes then pulled in to examine a rapid. It looked easy and was.

Next stop was Gypsy's Bane. Here, on an earlier expedition, Robbie Paul, the gypsy for whom the spot was named, flipped and for a long time was presumed lost.

This rapid has a large pool and drop construction, very much like the biggest rapids on the Colorado and quite unlike most of the rapids on the Omo. The sound of Gypsy's was also familiar from Colorado days, a deep booming roar providing base to the treble hiss of rushing water.

## THE TREK CONTINUES

After the morning's experience there were a few grins and lugubrious jokes as we drifted in to the landing, but the underlying tone was serious. A long examination of the rapid ensued. Jim Slade's boat was selected to lead. I remained in his boat. Clarice and Murphy, spooked by the morning's upset, decided to walk the rapid and take pictures of the run. Lou took Murphy's place in our boat. Meph, a tall, strong, California gal, raised on a surfboard, took Clarice's place.

We positioned on the tongue of the rapid and started our slide. Down we went.

Immediately everything went wrong. The chosen route was initially down the right side, to avoid a large projecting rock and a downstream hole on river left. This starting maneuver to the right was to be followed by a sharp swing to the left to skirt a large and dangerous hole on river right. Then we had to move back to river right to clear a feared and even more dangerous hydraulic on the left.

A hydraulic involves a sluice of symmetrical configuration over a large submerged rock. The geometry is such that a horizontal vortex is created downstream of the submerged obstacle. This vortex can hold a boat in place indefinitely and people can be sucked down and held under water for several minutes. The picturesque description of this often fatal condition is *caught in the washing machine.* The only way out for a swimmer is usually to dive deep enough to get completely under the vortex so that the river current will sweep you clear, not easy with the flotation gear that we wear.

As Slade took us down the tongue he was stroking hard right, but a slight quirk of the river, or maybe a misjudgment, pulled us left. Suddenly I could feel a powerful acceleration and the boat took off as if it were propelled by a giant sling shot. Down we went, careening first into the left side rock we had been trying to avoid. As we bounced off Jim popped an oar and we were totally out of control. By the time he got the oar back onto the thole pin it was too late. We punched through the hole below the big rock okay, but immediately downstream of that was the hydraulic.

In we went. Everything stopped. The boat was canted up about 45 degrees on the hydraulic wave and shuddering back and forth in that position. We tried climbing to the high side but things were very shaky and water was sluicing in and out from all sides.

I was hanging on by a single line when a wave picked me up and flung me a substantial distance upstream. As I was swept back towards the raft I had the impression that the boat was crashing down on me. I was able to get my feet up to fend off. With a powerful kick I pushed as deep as could into the depths of the river. The immense turbulence carried me down further, far underwater. The water was very dark down there.

At this point my ocean skills took over. I relaxed into a ball to conserve breath and trusted the buoyancy of my Mae West to eventually bring me back to the surface. I wasn't too worried because in these circumstances I can hold my breath three to four minutes without losing consciousness.

I really wasn't down that long. Probably only fifteen seconds or so. To the people on shore it seemed much longer.

I still had one hole to go through, however. Murphy, standing on a rock, alerted me to what was coming and I started paddling furiously to clear myself to one side. Murphy kept screaming "feet downstream." At the last moment I swung around into the preferred position and punched on through. After that it was a matter of getting hauled out.

## THE TREK CONTINUES

Lou was flipped out of our boat next. Murphy said he got a picture of Lou in the middle of a back somersault. Meph hung onto the boat gamely for a while, but was flung off and left behind in the hydraulic when Jim levered the boat out of the hole. Her predicament was particularly dangerous, and had everyone seriously worried. Jim plucked Lou out of the water and then grabbed me. But Meph was nowhere to be seen. Two other boats came by, but no one had caught sight of her. This was looking more and more like a major disaster. After we landed, gloom descended.

But not for long. Soon, merrily rowing along, came the final boat. Meph was sitting tall on the duffle. We all breathed a sigh of relief. Meph looked puzzled at our concern. She claimed that she really had had no trouble; she just went for a swim, that's all. I didn't believe her, and apparently neither did anyone else. For a while the River God had toyed with her life. Fortunately, the God grew tired of this game and spit her out.

There were already so many features in the Omo with Jim Slade's name attached, this hole was dubbed *Head Honcho Hydraulic*.

## DOUBLE TROUBLE—THE OMO

*Notes*

1. I did manage to lose my hat in the melee. A serious business on the river. Now I would have to use a bandanna to protect my head from the sun.
2. A check of our traveling library (which fortunately survived the mishaps) told us that adolescent crocodiles, in order to avoid being eaten by their suddenly cannibal elders, swim away from the colony and dwell under the rocks near rapids until they are grownup enough to return home. So I was no longer considered nuts. This new information would undoubtedly change our swimming habits. And, I might add, the crocodile's archaic solution to problem children probably should be examined more closely as a guide to dealing with human teenagers.

# MYTHOLOGY

WHAT HAVE WE MODERNS LOST? *We listen to the myths and legends from ancient times and react with disbelief and scorn — how could the ancients have had such silly ideas? Just imagine, they believed in Djinns occasionally trapped in bottles; they believed in Fauns, such as Pan playing his pipes. What nonsense!*

*Well, I'm here to tell you that Djinns and Fauns are real. I know this because I have witnessed both a living Djinn and Pan, himself, in the wilds of East Africa.*

*It was long ago, decades ago. Our Omo River expedition was camped on a sandbar. Supper was over and I had retreated to my tent in the growing darkness to relax, write and read by the light of a small camp lantern.*

AS USUAL, MY TENT OFFERED SHELTER from biting river insects. This night was different, though. Insects beat against the walls of the tent with unusual intensity. They wanted in. They wanted my light. These were huge insects, not the small biting kind. As best I could, I ignored the persistent, rattling, rain-like drumbeat. It grew darker.

After a time, one of my companions came to my tent and said that I should come out, there was something I just *had* to see. Struggling out of the tent entrance, I looked up and lo, far off, was a vision most magical and majestical.

# MYTHOLOGY

It really was — magical and majestical. From the far end of the sand bar rose a writhing, whirling funnel of light, twisting up and up, a luminous, dancing tornado, until it vanished in the darkness. My thoughts were drawn to stories of Scheherazade and the Arabian Nights, tales about Djinns and their magic and wishes fulfilled. And there, rising in front of me, was a veritable Djinn.

This I had to see up close. As I walked toward the phenomenon, the luminous funnel gradually resolved itself into millions of insects whirling in a frantic erotic dance around the light of the campfire. The insects were giant mayflies. Once a year they gather for one night of frenetic mating before death sweeps them away. *This* was the night. The nearby mayflies, rising from the river at nightfall, swarmed to the light.

That part of Africa is *dark*. Tsetse flies — the local species of those little monsters — kill cattle. The natives can't survive without their cattle so they avoid this entire region. Since ours was the only light for who knows how far, we were inundated. Pheromones released from our local mayflies wafted away with the breeze and enticed mayflies in from dozens of miles along the river. Here was the light, here was the place to make love.

As the flittering creatures whipped too near the leaping flames, wings were singed and bodies fluttered slowly to the ground. Within a few minutes the area, for several feet around the campfire, was covered in a carpet of writhing bodies several inches deep. The scene was straight out of Dante's Inferno.

The whirling frenzy continued long into the night.

In the morning I returned to a scene of desolation. Around the smoldering remains of the campfire was an enormous

## THE TREK CONTINUES

ring of mayfly carcasses piled high. It was grim — death in the millions. The Djinn had vanished. Though he granted no wishes, he left us, perhaps in compensation, the memory of his enchanted, deadly dance.

Pan was different. Pan I did not see, for Pan was shy. Pan I *heard*.

Coming from civilization, I had a lot to learn. Jungles are *noisy!* At least in this part of the world they are. The racket can be overwhelming. Birds sing songs, animals roar in the night, monkeys chitter and howl in the trees, insects rasp out their instrumentals. But mostly there are frogs. Little frogs, so small they disappear under the tiniest of leaves. Big frogs you never see because they hide in the reeds and bushes. Colorful frogs, frogs you only know about from hearsay. Noisy frogs. Always noisy frogs. Frogs with deep voices. Frogs with high pitched voices. Frogs that peep, peep, peep, or luuu, luuu, luuu. Frogs that HUM like a high tension power line. Frogs without number. The jungle is an endless cacophony of randomly mixed, excruciatingly loud noises — mostly made by frogs.

We had traveled much further down the river from our Djinn experience when I encountered Pan. The river had finally flowed out of its deep canyon in the mountains and into a veldt region known as the "Journey of Death" from an ill-fated early expedition. We were in a northern finger of the Great Rift Valley, some distance from the Omo's destination, the great Lake Turkana, or Lake Rudolf as it was formerly named.

We camped on river left. The beach nested below a low, flood carved bluff. Most of the area was barren, but upstream, some distance, was the edge of the riverine jungle. Because of the arid nature of the region the jungle extends only as far from the river as the roots of the vegetation can find moisture. Here it was confined by the bluff to a narrow area just north of the sandy beach.

Tranquility is a word that might describe our camp after sundown, with its scattered array of tents softly illuminated by a cloud-checkered sky. This peaceful scene belied the greater hazards — people hazards — as we moved from the wild mountains to the more settled lowland.

Sometime after we set up camp that evening a group of Bodi tribesmen materialized on the bluff above us. Silhouetted against the darkening sky, their lion spears vertical, the line of tribesmen created a primeval apparition of menace. Fortunately, they did not come down.

"THE DARK CURTAIN OF NIGHT fell swiftly with the setting of the sun." As in, "Who turned out the light?" That hoary literary conceit about the tropics is simply untrue. Admittedly, the equatorial evening is relatively short when measured against the long, lingering twilight of northern climes. But, here, only a few degrees above the Equator, cool evening lasts long enough for a pleasant and relaxed interlude before full darkness, with its strange stars, settles in.

Evening is the time for bathing — a delicious refreshment after a long day sizzling from sun and river-reflected sun. But bathing along the lower Omo is not a casual affair — there are crocodiles here. And crocs on the Omo are nothing like the crocodiles or alligators found in other parts of the world. These Nile crocodiles are enormous, ferocious monsters. Ferocious, aggressive and lightning quick.

Each beach along the river typically has a protected area which is suitable for bathing. In these places the swift current of the river is tamed by a quiet backwater eddy. These are the places where gazelles and other land dwellers come to drink. These are also the places where the crocs hang out, waiting for that brief moment of inattention when they can hurl themselves out of the water and snatch an unwary victim. For those humans living near the river there is a one in three chance that a crocodile will in fact make a meal of them.

Because of this situation, there is a well-tested procedure for human bathing in this part of the world. First, one never leaves the tent area without a companion. You pair up with anyone who happens to be handy. One person stands guard while the other does what needs to be done. To prepare for bathing you first gather a substantial pile of large and medium size rocks: *Croc Rocks*. Before entering the water a survey of the area is called for. Sure enough, we saw the paired bubbles in the water which said that crocs were eyeing us. We fired several smaller croc rocks at the various croc eyeballs, followed by the larger rocks. Kerplunk! The calm eddy temporarily cleared, for a while it would be safe to bathe. Periodically, your intrepid, and alert, sentry will fling more croc rocks to ward off the stealthily stalking reptiles. Bathing completed, the roles of overwatch and bather are reversed.

At this camp the bathing pool was upstream of the campsite at the southern margin of the riverine jungle. I went there to bathe. And it was there I heard Pan play his melody.

All afternoon the moisture-heavy monsoon, blowing in from the Indian Ocean, had piled thunderheads high above the mountain massif that had been our home for weeks. This evening the riverine jungle was making its usual unbearable racket. Suddenly a flash of dazzling light illuminated the growing dimness of late twilight. BOOM, CRASH, came the immediate thunder. That strike was close! The jungle went silent — dead silent.

The silence continued, and continued, until at last a single musical note, *luuu*, emanated from the green. The *luuu* was oddly soft and flutelike. The liquid sound repeated, to be followed, a second or two later, by a different *luuu* — a different tone a half step higher in pitch. The two notes coupled together and repeated. Then a third note, again a half step higher, attached itself to the second. As the scale repeated, more notes assembled onto its end. Tentatively at first, then with increasing confidence, more *luuu's* linked themselves — each a half tone higher — onto the growing chromatic scale. Within minutes the creatures of the forest had assembled a complete set of panpipes. Sliding up the chromatic scale was the breathy voice of Pan himself.

Astonishing! The ascending scale was one familiar to western ears. This was followed by still another surprise, for after a few ascending arpeggios, the sequence reversed and Pan slid down the panpipes in a descending chromatic arpeggio. Then, to top it all off, the sequence of tones went up and down, alternating in waves. Pan was having fun with his visitor — seducing him with his simple melodies.

After a time the rain ceased. Pan grew weary of his seduction and the carefully assembled pipes disintegrated into random discord. Less melodious voices joined in. The racket of the jungle had returned.

*A Flying Beetle, yellow and black. Photo taken by author near the Omo River*

# BUGS

BEFORE WE LEAVE THE OMO ADVENTURE, let me tell you more about the bugs.

*The canyon of the Omo River is uninhabited and was, before its recent dam construction, pristine. This is because people in the area survive on cattle and the variety of Tsetse Fly found there carries a lethal cattle disease. These creatures are a most unpleasant nuisance for humans but the diseases they carry are not fatal in this part of the world, unlike in other parts of Africa.*

**From My Journal**

OMO RIVER TSETSE FLIES are incredible creatures. If anything they exceed their reputations for painful bites. The day begins with dawn raids and the flies continue pestering until the sun goes down. They swarm thickest where the hippos hang out, but harass us everywhere else along the river, and off, as well. During the heat of the day, which around here means whenever the sun is up, they become ferocious.

They are unbelievably tough. Most of them are about a half inch long. Slap them, crush them, grind them underfoot. You will be convinced they are dead. After about a minute the corpse will twitch a bit, then rise, shake itself out and fly back to the attack. It is like something out of a horror movie. About the only thing that really works is to pull the beast's wings off. That way at least they can't fly.

Most everyone on this expedition is an environmentalist of one creed or another. The motto going in was "take only photos, leave only footprints." Tsetse's have engendered substantial behavior modification, however. Now, even the most environmentally dedicated can be seen to frequently trap a Tsetse under a cupped hand. The monster's wings are plucked off and the remaining, squirming, body is flung into the river with a malicious grin and the exclamation: "Drown, you son-of-a-bitch!" Much applause follows.

*My journal records some practical experience in dealing with Tsetse flies.*

If you leave any skin uncovered they will find it, especially if it is in an inaccessible location — like *behind* your armpits. Try reaching there! They will bite through two layers of clothing and even heavy wool hiking socks.

Dive in through the loose flaps of your tent and they dive right in with you. Quick devils. There you are trapped inside your tent with a half dozen Tsetses. But there is a solution: The big game hunt. Out comes the sierra cup and you're off. Chase about the tent until one of them settles on the wall, then quick, slam the cup over him. Run the cup along the wall until you've got him at a small hole at the entrance flaps. He'll tumble out and fly off, dazed. It takes a few minutes to clear a tent, but it is well worth it for a bit of peace.

*Tsetse flies are not the only bug along the river.*

Mosquitoes come out after dark. They are unobtrusive and it is rare that you are aware of being bitten. However, after a while a new crop of welts informs you of their presence.

# BUGS

In the time between the Tsetses of Bedtime and the Onslaught of the Mosquitoes there is a brief respite, when one can bathe in the river with a fair degree of comfort. This lasts fifteen to thirty minutes.

The only problem in that interval is with *Simuli* flies. These are tiny biting flies related to No-See-Ums. They don't seem to bother me much but others on this trip react with large welts.

It turns out Tsetse flies cause allergic reactions in some people. Murphy's back is a solid mass of hives — perhaps hundreds of them. I'm lucky in that I don't seem to react to them — yet! Except they do get through my armor at specific places, and multiple piles of bites ultimately lead to tenderness.

I don't want to exaggerate the problem, however. The insects are enough of a nuisance that permanent residence here would be distinctly unpleasant. However, for the short term of this trip they are more an annoyance and topic of conversation than anything else.

Other insects are more benign, and in many ways, more interesting.

Ants range from very tiny and multitudinous to very large and few. On a hike up into a side canyon there were large stretches of forest in which the ground and trees were literally furry with ants scurrying in every possible direction — paths crossing at random.

At one location, along the trail, the various scurrying ants formed a carpet which was several yards wide and so thick that the ground could not be seen beneath the living fabric. We had no choice but to step on top of this carpet. Oddly enough, the ants we were treading on left us alone. Huzzah! We had no ants crawling up our legs.

## THE TREK CONTINUES

At another location a small stream and massive horizontal tree branches blocked further progress on the ground. Like Tarzan we swung up into the trees to gain passage. One of the horizontal tree branches was a major highway for ants. Ants of many different species and sizes were running back and forth along the length of the branch, crawling over and under each other, without any conflict. The ants on this highway formed a kind of thick crest, inches in height and a colorful mix of oranges, reds, and yellows — much like the crest on an old Greek helmet.

The high temperatures and uniform daylight hours promote very rapid chemical and biological decomposition and ants flourish in a way I've never seen at home.

Other insects are different, as well. Many of them are similar to those at home but here they are very large. For example, the other day I saw a fly crawling on the boat which looked just like an ordinary housefly, except that it must have been at least two inches long!

Spiders also grow big. House spiders look more like large tarantulas than the tiny skittery things we have at home. We felt there was little or no danger in these dinosaurian insects as they are probably not venomous, and in any case, are quite sluggish. Some of them are very beautiful, too.

Butterflies we see in abundance, marked with quite lovely and unfamiliar patterns — although the patterns are not the iridescent glories that we associate with tropical jungles. In general it is too dry along this river for jungle insects to flourish.

Dragonflies here are a joy. Their colors are various shades of blue and aquamarine, and warm colors like gold, orange and red.

A few days ago we were sitting by a cascade in a side stream having lunch and lazing. I had the opportunity to observe a blue dragonfly for a considerable time. He had a favorite rock in the center of a small pool. There he would sit, usually with his wings displaced forward and down and his tail raised slightly in the air. When another dragonfly wandered into his vicinity he would fly off to engage it, chasing it away. I suddenly realized that this insect was behaving in what appeared to be a classically territorial way.

Another possibility was that the beastie was trying to mate with the passing dragonflies. Continued observation revealed a variation in the behavior pattern. On one occasion another dragonfly landed on the rock and The Pondmaster, instead of chasing this newcomer away, changed to a more elaborate display. Was he trying to attract this newcomer as a mate? After a while the intruder flew off and the owner of the rock seemed a bit deflated for a time. A few minutes later, another passing dragonfly appeared. This time The Pondmaster chased the new arrival out of the sky and things settled back to normal.

Since territorial displays are usually associated with mating duties it seems likely that the dragonfly was exhibiting both aspects of this compound phenomenon: preserving territory and the mating dance. It is fascinating to find supposedly vertebrate behavior patterns in such a very distant arthropod relative. This shows just how deeply ingrained territoriality must be in our biological heritage. It seems incredible, with evidence as fundamental as this, that people still seriously deny that man has territorial instinct — but many do.

*As all things must, our time on the Omo River was soon over. We were picked up by Carl-Gustav Forsmark, an East African legend, and taken up to his establishment in two trips of his old Russian microbus. There we would get air transport back to Addis.*

Camp here is by the Mui River, which is infested with Bilharzia, otherwise known as Schistosomiasis. Thus, all water here *must* be boiled and there will be no bathing. Serious business!

More insects are around us, strange and fascinating. Today I saw a four inch long walking stick, a three inch long wasp, a very large iridescent green and blue fly, a beautiful flying beetle, all gold and black, and a variety of brilliantly colored tropical butterflies and moths. These jewels of the tropics are among the marvels I came hoping to see.

We are camped, this evening, beneath tall trees. A colony of black and white Colobus monkeys chatter and leap back and forth in the trees' tops. Not particularly disturbed by our presence, they have elected to stay right here, rather than move to more distant trees. This is the first time we have been able to see Colobus close up. They are most impressive, most refined, decked out in their tuxes.

We're all enjoying our evening of rest, but as I try to get up to go to the fire, I'm in for a rude shock. Blisters on my feet, caused by a biting creature, have finally incapacitated me. These blisters have been developing for a week but, by late today, things clearly got bad enough that I am now temporarily lamed.

Carl examined my feet, shook his head at what he saw, and dug a large spray can of insecticide out of his duffle. Quite generously he insisted, over my protests, that I use the *whole* can to fumigate my wool socks. The whole can would be needed, he said. This was a considerable sacrifice on his part as the insecticide had been imported from America at great expense. Sure enough, upon liberal application of the deadly potion, a tiny, brilliantly crimson, little tick crawled out of one sock and expired on the floor of my tent. This minute creature explained my suffering. The bites had started out on one ankle, but when I put my socks on in the morning they went on different feet, so both feet were exposed to this bloodthirsty predator. Repeated bites and the vesicles merged until all but the soles of my feet had become one giant blister — spreading to well above the ankles.

Doc Walt said he was going to have to operate if I was to regain any mobility. So surgery by the campfire became a significant part of my evening's entertainment. Expedition leader Jim Slade chased the other members of our crew away. He may have anticipated that I would embarrass myself with yowls of pain from surgery without anesthetic. Fortunately, that did not happen.

Walt prepared his instruments and sterilized them in the flames while his wife, nurse Bretta, made up compresses and bandages. Then the two of them set to work — slicing and dicing. Oddly enough there was little pain from the surgery. The wee creature had done enough damage that lancing the blisters actually produced relief. Salved and bandaged, and with promise of no more biting, I immediately felt much better.

*That little scarlet tick infected me with the often fatal Relapsing Fever, a cousin of Lyme Disease. I suffer from it still, but that is a souvenir which I have accepted as the cost of good memories that I wouldn't trade for anything.*

# THE FAIRY

*"You misjudge me," he must have thought as he glanced in my direction. "You misjudge me, so let's have some fun with this," his thinking likely continued.*

I WAS SITTING TOWARDS THE REAR of a Greyhound Bus. The bus had been trudging slowly up the road from the center of the fishing port of Haines, Alaska. In that remote part of the world I believed that we, our party, had the bus to ourselves, for few traveled this route. I was one of a group of river runners and professional boatmen on our way to travel the Tatshenshini River, from Dalton Post in the Yukon Territory, across British Columbia, across inland Alaska north of the panhandle, to meet up with the mighty Alsek River and thence to the sea. The boatmen were particularly excited, this day, for the famous writer, Ed Abbey, was to be one of us on our expedition.

At some intersection of the town we stopped to pick up a couple of newcomers. The first to board was Abbey. Tall and stately, sporting a luxuriant beard, he stepped aboard the bus in all his regal majesty and took the lead seat that had been specially reserved for him.

Following behind Abbey, looking somewhat confused, was some kind of street bum. Small, disheveled, weather beaten, stubbly red faced, snaggle-toothed — those few teeth

# THE FAIRY

that remained. He was arrayed in an old checkered lumber jacket, faded shirt and levis worn to near rags. A glance at this apparition and I quickly looked down at my book, hoping against hope that he would not notice me and decide to take the seat next to me. All around were empty seats for him to take.

My hopes were dashed when he staggered down the aisle to where I was sitting, stopped, grinned maliciously and said with a slurred voice, "Is this seat taken?" Since it obviously was not I had no choice but to invite him to sit.

The derelict leaned close to me, his face only inches from mine. His jaw hung open. "What do you do?" he mumbled, indistinctly. "I'm an engineer," I replied, backing away a bit from his intrusion. "Why are you here?" he asked further, leaning forward a bit more. "We are on our way to run a river." "What river?" he importuned, moving even closer. "The Tatshenshini," I responded, my voice getting cold as ice. By now I was snugged up against the equally cold window of the bus. The bum just kept looking at me, not saying anything. Finally, knowing I was being manipulated, I asked in return, "How about you, what do you do?

A great slack-jawed grin broke out over his face as he contemplated his answer. He moved even closer so that I could hear the soft wheeze of his lungs and feel his breath hot upon my face. Then his voice slurred out at a somewhat higher pitch: "Oh.... I'm a *fairy*..... I'm *a fairy*, and I'm going to be with *you* for *a long time*...."

Involuntarily, I cringed.

He paused and moved back a bit. His face changed, becoming normal, his voice cleared up, strong and resonant, with a normal cadence and a laughing lilt, "I ferry boats, I ferry planes, I ferry trucks. In fact, I'm to ferry your boat truck back to Haines."

He pulled all the way back, chuckling softly, having had immense fun with my discomfort. I let out an involuntary sigh of relief. My new acquaintance moved across the aisle to the empty seat there, out of politeness and so that we both could have more room to stretch out.

It was evident that, however he may have looked, here was a man of considerable intelligence and wit. Our conversation resumed. This time I was really curious about my new companion. I inquired further about his occupation.

"Most of my life I was a cowboy, I had a large spread in Colorado. About ten thousand acres." He paused. I was impressed. "That's where I lost my teeth. Cowboying is a tough, rough life. You get banged up a bit. After the kids moved away and my wife died, I had no real reason to keep up the ranch, so I sold it and started a new life."

Having spent his life outdoors it was natural that he would seek an alternative that would keep him outside and in the wilderness. With the proceeds from the sale of the ranch he could do pretty much what he wanted now. Part of the time he worked as a *ferry*, moving trucks, boats and planes around from one part of this big country to another. Much of the rest of the time he fished the open waters of the Gulf of Alaska, fishing even in winter.

In running his ranch he had developed many skills. He could fly several types of aircraft, though his helicopter was, all around, his most useful tool. With it, he could herd cattle down from the high country without the pain of long hours and days in the saddle. He also could operate just about any kind of heavy machinery. On a ranch one did not often hire a contractor, a rancher got used to doing the job himself. Large boats he learned after he left the ranch, becoming qualified both as skipper and harbor pilot.

I commented on the unfamiliar shape of the boats in the fishing fleet back in Haines. I was acquainted with deep sea fishing boats, having spent a week at sea aboard a Tuna Clipper. These Alaska boats were much more strongly built than the ones I'd known; many had rounded, reinforced sterns and considerable tumblehome. My new friend regaled me with stories of the sudden storms and mountainous seas that he had encountered in the Gulf of Alaska. The Haines boats could ride these dangerous waters as safe as ducks, which they greatly resembled.

We talked of planes and what it was like to be a bush pilot. He told of the hazards of getting lost, of the fear of rapidly changing weather, especially up around Anchorage, the pivot point for weather systems, where the weather could be clear skies and sunny at take-off and socked-in, completely blind flying a few minutes later. I mentioned that a new satellite navigation system, the Global Positioning System, was finishing its development and would soon start its deployment. GPS would tell you where you were within a few meters, anywhere on or above the Earth. He wanted it. He wanted it very, very much. I explained that GPS receivers would be extremely expensive in the beginning — the technology was very fancy. He replied that it did not matter how much they cost. In this part of the world they were worth every penny and would be a hot selling product.

"You flew up here from Seattle, didn't you?" I told him I had. I flew into Juneau then was a deck passenger on the Mail Steamer up the Inland Passage to Haines. "What was the country like that you flew over?" I remembered a thousand miles of trackless, primeval forest — no roads, no towns, no settlements, no humans of any kind — for endless hours. The old rancher said he flew that route regularly, moving aircraft back and forth between various parts of Alaska and Vancouver or Seattle. The more experience he had over that route the more frightening it became. "Imagine if I had to make a forced landing somewhere in that damned forest. Where would I be, what direction should I hike to get to people? What chance of survival would I have unless I knew where I was? Or, suppose I was out at sea, off a rocky coast, and was caught in one of those horrible fogs we have around here? You bet I would pay *anything* for your GPS! I'm gonna get me one just as soon as they come out."

It is a hundred and fifty miles from Haines to Dalton post, a hundred and fifty *slow* miles over the winter-resistant gravel road. We had been traveling up the Chilkat Inlet and had finally passed beyond the estuary of the Chilkat River. This broad river valley is the breeding ground for many of Alaska's Bald Eagles. High in the trees along the roadway perched the silver headed eagles, hundreds of them. Hundreds more whirled in the air further down in the river valley fishing for the fat salmon cruising up the river. It was high summer here, the time for feasting, nesting and frivolity.

My new friend told me stories of ranch life and the trials and tribulations of punching cows, of his struggle maintaining the health of the cattle through long, bitter winters, of the hazards of roundup and branding, of the pleasures of country festivals and rodeos. Though he

# THE FAIRY

claimed otherwise, I felt that he missed the cowboy's life, hard as it had been.

We left the river valley and climbed over a range of mountains, the gravel road crunching under our tires. I expected that we would descend into a similar valley on the other side, but I found we were driving across a high country of rolling hills, mostly above the tree line. Occasionally, in the far distance, snow-capped and glacier-flanked peaks became visible above the barren, green grass countryside.

The landscape appeared deserted, but even here there were settlers. We came upon a small valley that meandered away to the right of the road. My companion told of an old gold miner who lived far up that valley. The miner earned a surprisingly good living panning gold from the nearby stream. A century of neglect after the great gold rush had replenished the gold in the stream bed. My new friend stayed with the miner one winter. It was fortunate that he did, too, for the old man grew sick and it was the extraordinary efforts of his guest that saved his life.

The gravel road dipped down to cross a stream over a narrow, one lane bridge. The bus slowed to a crawl as it approached the bridge. My grizzled companion started cackling. It was here, at this bridge, that he had fought a mighty joust. One day he was bringing a big tanker truck south to Haines. As he started to cross the bridge he was confronted by a big bull moose standing solidly at the far end of the bridge and blocking his way. He stopped the truck at the entrance to the bridge. Tired after a long drive, he was not inclined to be patient, so he honked his horn, geared all the way down and started slowly to move out, planning to shove the recalcitrant animal aside. Instead of retreating the old bull just lowered his head and charged.

The truck didn't yield from the collision, but the radiator did. Pressurized Prestone sprayed out every which way! The moose was startled by the truck's refusal to budge. Twice more he charged before giving up and wandering off, leaving behind the wounded and stranded behemoth. How did he get out of it? He didn't say.

We traded tales and laughed our way across the land.

At last we reached the junction to the steep and precarious dirt road leading down to Dalton Post and the river. The boat truck had preceded us and was waiting for us to arrive. The boatmen climbed off the bus, grabbed their duffle and quickly unpacked some items. Within minutes all of them had strapped on holsters filled with heavy pistols, 44 magnums being the favorites. A shotgun was also broken out and kept handy as we walked behind the truck. We were in bear country now, grizzly bear country, and no chances were to be taken — especially after a dangerous encounter at this very spot the previous year.

Our ferry drove the boat truck down the hazardous dirt trail to the river. We all pitched in unloading the truck, for mid-afternoon was upon us, and although these high latitude days were very long, the truck had to be taken back to Haines, over that long gravel road, before the short night had grown fully dark. My friend climbed aboard the truck and we shook hands for the last time. I knew that I would never see him again, but these hours together were a treasure that I would carry in my memory.

## THE FAIRY

I did become acquainted with Ed Abbey. Spending weeks in the wilderness together — sharing meals, climbing mountains, traversing glaciers, enduring storms with someone, you do become acquainted, for better or for worse.

Abbey certainly was a man of accomplishment and could tell a tale with the best. Still, I came away from the excursion with the sense that my remembered friend, *The Ferry*, had lived a far richer life.

*Avalon Harbor, Catalina Island*

# SEA STORY

HAVE YOU EVER HEARD A SCREAM? I don't mean a fake Hollywood movie scream. I mean the real thing — a blood curdling, hair raising, gut twisting howl of astonishment and terror. Well, I have. I heard it late one night on a distant island.

I'll start at the end. It was 1972. This was a long time ago. A time when people were worried about the Cold War. But on this trip, that was a very distant concern. Picture a gently undulating mirror that was the sea, a sea that quietly reflected the starry glory of a soft summer night. The ocean here was nestled within the protecting channel between Santa Rosa and Santa Cruz Islands. In the far distance was a sprinkle of dim sparks along the Santa Barbara coast. All around us were other boats — a multitude of boats, their lights dancing slowly in the low swells. Sweet accordion harmonies and masculine songs enveloped us from all sides — we were anchored in the midst of the myriad vessels of the Russian fishing fleet.

For us this, too, was a night of celebration — it was our last night at sea. Sometime after midnight we would start the long haul east and south to our final berth at Dana Point, and then home. The morning would have its charms as well as we cruised along, escorted by a large pod of gray whales, and a school of dolphins playing happily in our bow waves, wishing us bon voyage and good cheer.

## THE TREK CONTINUES

But that was the still unknown morrow. Tonight we celebrated as all travelers on the sea celebrate their night before landfall. We sang songs, we laughed through skits and we told stories of our recent adventures.

We talked about that day's events, a day spent roaming the uplands of San Miguel Island. A place now desolate, but once teeming with human activity. Think of the ghosts of San Miguel. Think of the Spanish explorers, captained by Juan Cabrillo, now only a ghost himself, still haunting the living, his bones buried somewhere unknown, within the bleak landscape of this island, or maybe that. Go back further. Go back long before the Spaniards, before the Mayans, back even before history, indeed, back all the way into the last ice age.

It is said that crowds of people, large crowds, through untold ages, earned themselves a good living on this island. We call them Indians. They, no doubt, called themselves "The People," as do most tribes on earth. The People came to San Miguel because it was a place of riches. And, maybe in that distant time San Miguel was verdant like its nearby neighbors. But probably not, for the west wind ceaselessly whips in from the boundless Pacific Ocean, beating down the vegetation, scouring the flat-topped island. There is no shelter here from the wind. But still, this island is surrounded by the infinite bounty of the sea.

If the land is bleak, the setting remains magnificent. Northside, a long crescent bay, with a broad white sand beach, curls around a small island set in crystal blue water. On the south and west, high cliffs plummet down to rock-guarded inlets pounded with high tossed sprays by the long, powerful Pacific swells. Sea lions, elephant seals, and birds love the south and west. The west end of the crescent bay has a high cliff as well. This is the area where The People lived and worked.

Entry onto the island had not been easy. Our boat anchored well offshore and we rode in to the beach in an inflated Zodiac. We waited until the entire group had been shuttled ashore. There had been no way to avoid the wet of the landing, so we had to walk our shoes dry as we climbed the narrow trail up to the plateau.

Our task was to learn, and learn we did. The ecosystems changed dramatically during a climb of only a few hundred feet. Salt-resistant succulents gave way to waxy-skinned, drought-hardy bushes and, of course, cactus, and then to withered grasslands and strange shrubs on the plateau. As we climbed we learned the zonal transitions and the species of both plants and small animals.

On top there was not much to see, not at first, anyway. Near the dilapidated wreck of the old sheep rancher's home we found a small monument to Cabrillo. Trails radiated out in different directions. We took the one leading west.

Nothing caught our attention on this trail, just endless, weather-beaten, dried-up prairie. If many people had once lived here where had they been and why? And what did they do for a living? After about a mile we began to get answers. A long, low ridge appeared anomalously from the landscape. About fifteen feet high, it was tens of yards wide and extended for hundreds of yards to the west, till it dipped out of sight with the contour of the land and continued on. This was a midden, a dump made by countless generations of The People.

There were no organics visible on the midden, just dusty sand and stone chips. It is these flakes that tell the story. These were a special type of chert, a fine-grained silica laced rock which is particularly suited for tools — arrow heads, spear heads, scrapers, knives of all kinds, and much else. The midden had been the site of a long-lasting tool making industry, that was clear. But why was it here on San Miguel? And, where were the outcrops? Decades more of exploration would be needed before geologists finally found the source of what we could see then.

Why San Miguel? There were equally good sources of chert on the mainland. The answer lay in the combination of riches possessed by the island. Bitumen is plentiful from the oil seeps in the Santa Barbara channel. Abalone is commonplace and large-shelled in the waters immediately offshore. So is long-stranded kelp. Long ago some insightful wizard figured out how to combine those resources into something that distant people demanded. As far away as the Mississippi Valley archeologists find the San Miguel offering, sometimes still intact. It consists of the two halves of an abalone's shell, filled with an assortment of kelp-padded chert tools, the shells sealed together with bitumen to make a watertight package. Obviously this product had been traded from tribe to tribe, without being opened, until it reached a far corner of the Continent. The trading network was impressive. When did all this start? No one knows. But the midden is very large and obviously very old.

As we walked along the flat ridge of the midden something caught my attention. Next to my foot was a chalk white bone. Absently, I picked it up. After a moment I realized that this was not from an ordinary animal, it was human — a femur. I wandered over to our pathologist. He took the bone from me and talked about it as we hiked north along the peninsula that rimmed the west end of the crescent bay. Turning it this way and that, he concluded from the size, seams and structures that it had belonged to a woman of about five-and-a-half feet in height and thirty years of age. We kept the bone through this part of our excursion, intending to replace it when we had traced our way back to the midden.

On returning to the midden, I showed the pathologist where I had found the femur. To my surprise, there was the complete skeleton of the woman, laid out supine. Somehow I had completely overlooked what was now obvious. Gently we replaced the bone to properly complete the ancient remains. No doubt buried deep, her body had weathered out over the centuries. The midden was not just a place to work, it served other purposes as well.

Though we could not know then what we know today, San Miguel has been a magnet for archeological and geological discovery in the decades since our visit. And it turned out that the speculations of our leaders proved prescient. San Miguel was inhabited, at least twelve thousand years ago, by a highly sophisticated maritime culture. The artifacts they left — especially finely worked stone tools — show astonishing craftsmanship and highly specialized adaptation to the marine environment. It is even possible that people had been living there thousands of years before that date. We cannot know because as the ice age glaciers melted the sea rose — sixty meters (200 feet), at least, and people would have moved miles inland to the highlands. Today's island chain is a shrunken remnant of what once was a single, enormous island.

It was, as I recall, the summer of '72. We had packed 44 people, men and women, onto a 64 foot converted purse seine tuna fisher. Though we were on a fishing boat, this was an academic not commercial exercise — we were cruising the

Southern California Channel Islands in a hunt for knowledge. The excursion was a University of California course in Channel Island ecology. It was one of those things you sign up for never quite knowing how it will work out, but hoping for the best. In other words, it was *an adventure.*

People filtered onto the dock as the pre-dawn sky was brightening. We needed an early start for it was 45 sea miles from Dana Point to Two Harbors at the Isthmus on Catalina Island. Even leaving at sunup we would not arrive there until mid-afternoon and much was on the schedule at our destination. Two Harbors was the first stop on a voyage that would cover more than 400 miles and six islands.

As usual, getting started always took a while. As we boarded the vessel we were directed down a companionway at the aft end of the deckhouse. This led down into the bunk room deep in the hold of the boat. The walls of this room were lined with stacks of bunks, three high. Forward were a pair of heads, or lavatories. Topside there was an additional pair of heads in the deckhouse. These sparse facilities had to serve the large crowd. Patience and courtesy were to be essential. After getting our bunk assignments and stowing our gear we congregated on the aft deck. Then, a muster to make sure all were aboard, a short briefing and we were underway.

Dana Point Harbor is small and soon we began to feel the deep ocean swells. Although I had been to sea many times in these waters this was the first time on board a tuna boat. Actually, this was my first time *at sea* on board a tuna fisher.

I had, in fact, been on board one long before, when I was ten. My dad took a weekend day off from constructing the yard of our newly built tract home and we drove down to San Pedro for the annual Fisherman's Fiesta. This was a day of

# SEA STORY

celebration, when the Archbishop arrived from St. Vibiana's to bless the large fleet of tuna boats. As we walked toward the enormous mob gathered at the wharf, a young man came up, declared that we looked to be a nice family, and handed my father tickets to one of the boats. Naturally, Mom and Dad were skeptical about this gift, but decided to check it out anyway. Sure enough, the tickets were genuine and we were politely ushered aboard the already crowded boat.

Such a situation is always awkward because the people on the boat were family and we were strangers. When the folks there found out that Mom was Sicilian, that her father owned a fish market and that her grandfather had been part of the fishing community on the west coast of Sicily, why then we were indeed family. The fishermen of the fleet were Italian and Portuguese, and not strangers after all!

My lot was to sit with the other kids on top of the deckhouse where we had the best view. The fleet shoved off for the parade — our boat, along with the others, gaily festooned with flowers and flags. Periodically someone would climb the ladder up to us with a tray of goodies from the lavish buffet down below. We feasted and laughed and soaked in the sights and the cheers of the people on the land. It was a good day.

But this, now, wasn't the calm water of San Pedro Harbor. We were well out to sea and moving swiftly along. Long swells from some distant storm were marching up from the south. The shallow bottomed tuna boat rolled in a disconcerting corkscrew motion as we plowed our way north and west.

I began to get my sea legs. The deck rose and fell, slanting this way and that. Wait a bit and your destination would be downhill. An experienced mariner lets the vessel do the work. Downhill, always, is the preferred direction to walk. It is just a matter of waiting for a bit. Sailors stagger on board a ship? No, they are just being economical.

The boat's odd motion was getting to people. Some were starting to hang onto the rails. "Look to the horizon," I advised someone. "It will help." It did.

I am fortunate in not being prone to motion sickness. That green-faced monster is not a matter for amusement, though. I once did have a particularly bad experience of it — flat on my back in bed at home, my inner ears highly inflamed from the extraction of impacted wisdom teeth. Ultimate misery! Dramamine took care of it, then. Since that day I do not turn down the newer Bonine when it is offered.

After about an hour on the deep sea the aroma of frying bacon began wafting from the galley above the deckhouse. This set off a stampede for the rails among the more tender bellied. I went in the other direction — I was hungry. The galley stove and sideboard, on one end of the galley, extended laterally, completely across the upper deck house. Before it was a long counter with the usual raised edges to prevent plates from sliding off. The arrangement was like a compact coffee shop. Low, backless stools were bolted down to the deck in front of the counter. One would suppose that in a strongly pitching and rolling vessel it would be easy to slip off a stool. The opposite proved to be the case. Long experience had evolved this seating. Additional small tables, stools and benches were scattered around the small galley space. Quite a nice little sea going bistro it proved to be, and one that quickly became a favorite club house.

Breakfast was bountiful — bacon, sausages, eggs, hot cakes and much else. This first morning, though, not many people were occupying the seating. (That changed quickly as

# SEA STORY

people became used to the boat's motions. Within a couple of days one had to hustle to get a place in the nice, warm, cozy galley.)

The day wore on. It was a long passage from Dana Point to Catalina. I became acquainted with a few of my fellow passengers, students and teachers alike. Lunch time came, and more conversation. The sea was getting rougher and for the first time I began to feel a bit queasy. It caught me by surprise. Time to take a nap.

There is something magic about sleep on the sea. The rhythms of the vessel and the waves become embedded in one's body so that one becomes melded with the movement. It's like tuning a fine instrument into perfect harmony. This is an excellent cure for the queasiness that comes from the motion of the boat.

The sleep cure was something I had learned from my friend Pat Royce, author of the ubiquitous handbook, *Royce's Sailing Illustrated*. This was while heavy weather sailing, on a three day excursion, aboard his famous boat, the Pink Cloud. Why, I had asked, was the boat pink? Pat had fielded this question innumerable times before. "I teach sailing," he said. "The wives are almost always skittish when their husbands drag them down to the dock for their first lesson. When the ladies see the boat they melt and decide that this is going to be fun." Which it always is, he assured me. To fully appreciate the irony of this you'd have to envision Pat. He was a tall, ruggedly handsome, muscular man, well over six feet in height — not unlike John Wayne. And a devoted family man, as well. Not the sort you would expect to find sailing in a pink boat.

# THE TREK CONTINUES

## Catalina the Isthmus and Avalon

SURE ENOUGH, A COUPLE HOURS OF SLEEP and the boat's odd gyrations seemed to have almost melted away. By then the cliffs of Catalina loomed not that far off the port bow. Catalina is, for all practical purposes, two islands, joined by an isthmus, a narrow land bridge which is less than twenty feet above the mean sea level. We were heading towards this isthmus, which was still hidden behind a headland.

First stop was Fisherman's Cove, a short distance to the east of the isthmus. The University of Southern California had recently established a marine biology research laboratory here. This was to be our main interest for the rest of the day. After a tour of the various labs, and some sit-down lectures about the local marine life, all of us adjourned to the sand pit for a game of beach volleyball. Us against them. This was a rather too obvious attempt at team building. It didn't matter that it was obvious. It worked.

Since it was still some distance to the isthmus we tromped back to the dock and reboarded our boat. A slow passage close around the sea cliffs and the view of the settlement of Two Harbors, at the isthmus, opened up. The boat let us off at the Isthmus Cove dock then backed off to pick up a mooring in the anchorage. That night it would again be at the dock to pick us up.

The place had changed in the years since I had walked these paths as a boy. The old, bedraggled, clapboard cantina had morphed into a spiffy new patio restaurant. There probably was a new hotel, somewhere that I didn't see — there were a substantial number of people wandering around. This was quite different from the nearly deserted scene that I remembered. A few new homes were now widely scattered on the hillsides as well.

In the evening we joined the Marine Institute crew at the cantina for supper and laughter. Eventually enough was enough and we walked back to the dock and well-earned sleep. It is surprising how weary one becomes spending a day fighting the motions of a boat on the sea. Sometime after midnight the diesels started up and we moved out into the gentle oscillations of the now quiet ocean. Babes in a rocking cradle, we drifted back to sleep. Sleep became deeper.

Morning found us moored in Avalon Harbor. After a hearty breakfast we packed our sleeping bags and duffle bags and went ashore for two days of learning about these islands.

Solid land. It was rolling under our feet like the water we had just left behind. No earthquake, this, just the twenty-four hour tuning of our bodies to the sea — it typically takes a few minutes, or hours, once on shore for the land to steady down to solidity.

So it is that on land, sailors actually do stagger. After a long time on the water, their steps, like ours, are unsteady when they set foot on land. It is not unknown for even seasoned sailors to become violently seasick when they come ashore.

We stacked our duffle, or gear, on the shore near the pier. It would follow us, later. Our first goal: exploring Avalon.

In the story of Arthur, Avalon is a mythical isle of transcendent beauty. This Avalon was real, but it possessed some of that magic. It was early in the morning, and despite the sun sparkle on the water, and the growing warmth, the town was still half asleep. A couple of the restaurants were busy serving breakfast to the mariners from the boats anchored nearby. Only a few shops were open, though — their proprietors were visibly restocking and tidying up in anticipation of the multitudes from the mainland who would flood in later that morning.

Tempting as it was to linger a bit, and soak up more of the enchantment this scene radiated, we forged ahead to our first destination: the Casino, the famous rotunda which poked out from a promontory some distance to the west of the town. A museum in the basement of the Casino had the usual displays of dusty old odds and ends, dug up from various middens around the island. A small bronze Buddha instantly caught my attention. The legend printed next to it said it had been found at a pre-Columbian level in one of the middens, snuggled next to the skeleton of its presumed last owner. The bronze was well worn, indicating that it had been passed down through many generations until it reached its resting place. I had recently been studying Chinese art. The unique style of this small object strongly suggested that it had been made during the Song Dynasty and was perhaps a thousand years old. How it got to Catalina in pre-Columbian times is a matter for speculation, but clearly there had to have been some trans-Pacific communication or it would not be here.

Outside, waiting for us, was an old, long-nosed, yellow school bus. Our duffle was already piled up in the back. We climbed aboard and the bus, with a wheezing complaint, started the climb up the steep, twisty road leading from Avalon to the mountainous western interior of the island.

Soon we passed through a gate and into the restricted area of this Wrigley-family-owned island. The narrow, one lane road hugged the edge of the interior plateau as it wound its way around a succession of box canyons that plunged outward, down to the sea. The drive above the cliffs was slow, possibly because of the extreme hazard of driving fast, but really because our leaders were instructing us on the geology of the island, such as it was then known.

The bus stopped. We all piled out. There are things you cannot see by looking out a window. The road had curled around the inland edge of one of those steep canyons. Looking down from the edge could make you dizzy, the drop off was that precipitous. The canyon did not twist and turn, it was nearly a straight shot all the way from its head to the ocean. It was almost as if some Titan had swung a giant axe, and with a single chop, had created the canyon. Water erosion did the rest. Maybe the canyon broke apart along a fault line. That was the most plausible explanation.

At the bottom of the canyon was a line of bright green vegetation that told of water, even though the visible stream bed was dry in this summer season. Water was there, but below the surface. To get at it you would have to dig.

Despite the impressive scenery, the real reason for the stop was to explain a problem that plagued the island. At some time in the past someone had brought in pigs and goats. Perhaps it was the Spaniards. Perhaps it was the later gold miners. However it happened, enough of these animals had escaped captivity to breed a large population of feral descendants. These feral animals were more than a nuisance, they were destroying the natural habitat of the island. A program had been underway for several years to hunt these interlopers to extinction. As we now could see, this was not an easy task. Pigs could be lured with tasty bait and trapped or shot. Goats were another matter entirely. Goats were well adapted to these steep canyons, parts of which humans could not safely traverse. The goats, protected by the inaccessible terrain, were perfectly happy munching on the local plants, and by now had become extremely shy of people, so they could not be lured into ambushes.

After poking around the head of the canyon for a while we resumed our journey. Sometime after noon we arrived at the island's only airport, the so called "Airport in the Sky." I was curious to see this marvel. A number of my senior colleagues at work were pilots. Occasionally they would take a long lunch and fly over to Catalina to grab a Buffalo Burger. Sometimes they would bring back a hair-raising story about their encounter with this challenging place.

The airport contractors had carved away the top of one of the lesser mountains. The length of its runway was severely restricted by this fact. The runway peaked in the middle, obscuring sight, and at both ends the runway terminated in a steep drop off. Misjudge the landing and you would likely run right off the far end of the runway and drop into the canyon beyond. Misjudge the takeoff and you were in for a plummeting rollercoaster ride until you could gain sufficient airspeed to pull out. Enough wrecks litter the surrounding countryside to provide plenty of warning about the hazards facing pilots flying into or out of Catalina.

After a pit stop and lunch we resumed our trek. This time we were headed for the south side of the island. This was the Pacific Ocean side. Its flora was correspondingly different from the more protected north, or mainland, side.

First stop in the afternoon's journey was the Wrigley Ranch. The ranch had been established to raise thoroughbred horses. A wrangler greeted us, told us the history of the place, and gave us a guided tour of the corrals. The horses were magnificent.

# SEA STORY

The next stop was on a bluff overlooking the infinite Pacific — a living blue blanket, far below, that stretched to the horizon. This was a special place on the island, for it had a protected grove of California Coastal Live Oaks. These trees had long been familiar to me because of their proximity to where I had lived when growing up. But, other than climbing around their massive, low hanging, branches, I really did not know much about them. That was about to change.

The island goats were killing its oaks, we were told. How could a goat fatally injure such an imposing tree? They did it by eating the old leaves that had long since dropped to the ground. Live Oaks are messy. The dried-up leaves of yesteryear cover the ground around their trunks in deep piles. I knew about this from my childhood, but it never occurred to me that this debris was life giving.

As we know, Southern California is a desert. Yes, the area gets winter rains, but the rainfall total is small enough to classify the area as arid. Native trees, and other plants, have long since adapted to this relatively waterless climate. The Live Oak adaptation takes advantage of the morning coastal fogs which predominate from late spring through autumn. These fogs not only ameliorate the heat of the summer, they also water the oaks. The job of the leaf debris is to trap the morning moisture and convey it into the ground, where the tree's roots can drink it up.

This grove we were taken to was an experiment. In order to protect its trees the rangers had planted cactus around the periphery of the debris piles. Because of the trapped moisture, the prickly pear cactus rapidly multiplied to form a barbed-wire-like barrier which protected the leaf piles from intrusion. It worked and the trees thrived. The method was now to be deployed across the entire island.

There wasn't much left to the day. The bus trucked us back along the narrow paved road, took a turn onto a side road, and then, sometime later took another turn down a short stretch of dirt path. We had arrived at Black Jack Campground. Nestled between Black Jack Mountain and Mount Orizaba, this was a football field sized patch of goat-trimmed grassland gently sloping down in an easterly direction. Its primary virtues were its water supply, its park tables and, most important, its chemical toilets.

We disembarked, unloaded the bus, sorted out our duffle, and then started hunting for a place to bed down. Near the road was a buffalo wallow. Here, in this animal-carved sand box, the big beasts would flip themselves over and squirm around until they had shed the fleas and other insects which tormented them. We were directed to stay well away from the wallow in case any buffalo should wander by.

There is a debate as to how the buffalo got on the island. The prevailing opinion is that they were introduced for the filming of a silent western movie back in the twenties. However, that particular movie has no buffalo and was clearly not filmed on Catalina. Make a guess, it probably will be as good as any other.

The evening was spent in conversation around a delicious barbeque. With only gas lanterns to light the warm night, and draw in every insect still flying, we tired travelers turned in early. It only takes a little while before the hard ground seems almost as comfortable as a mattress. Fatigue helps the transition. My deep sleep was interrupted during the middle of the night by some kind of commotion at the other end of the campground. It didn't keep me awake long.

After breakfast, and some time to clean up, we reboarded the bus for the trip back to Avalon. Once back on the main road to town the bus stopped and we were booted off. No nice ride this day. Walk. It's only a bunch of miles! The rationale was that the right way to see things, and to learn about them, is to walk, not ride. The bus drove off, so walk we must, long miles, getting instruction along the way. Until, in a fit of mercy, we were allowed to climb back aboard the newly returned bus. Waiting for us inside were box lunches. These pacified us sufficiently that we were not too grumpy during the final drive to the harbor.

The rest of the afternoon was free. We checked out the shops, watched the tourists, and chatted or dozed under shade trees. Dinner was in one of the seaside restaurants. Then we found ourselves back on our boat. Best to get to bed early. After the day's long hike we were ready.

There is nothing quite like being rousted out of a nice cozy bunk after midnight to spend a couple of miserable hours benthic dredging the deep. We had been warned that this was going to happen at some indeterminate time in the future. Tonight was the night. When it happens, we were told, dress extra warm. Good advice. It was indeed cold, and clammy too. Since we were only very slowly underway, and well out to sea, the flat-bottomed boat wallowed in the glassy swells, making it hard to keep your footing on the dew-slick deck. It got worse. It also got interesting.

Why do this at night? Even at depths where sunlight barely penetrates, such as where we were probing, there is a diurnal cycle. In much of the water above the bottom the fish can see substantial daylight and behave accordingly: they eat during the day and sleep and poop at night. The nocturnal drift of debris to the bottom feeds a lively benthic (very deep) ecosystem. More importantly, night hides many bottom-dwelling creatures from predators so they can safely go about their business. Activity peaks around midnight. Thus, complain all you want, but get yourself up and on deck in the dark and the cold and the wet.

At the time, we were some three or four miles north of Catalina. Here the ocean depth is about half a mile. So our dredging would be deep indeed. The project's designers picked this night precisely because the ocean near Catalina is much deeper than it is near the other islands.

In an act of kindness our instructors had let us sleep through most of the dredging process. Half a mile is a long way down and it took a long time to unwind that length of stainless steel cable from the large drummed winch, and an equally long time to bring the dredge back to the surface. The process was almost complete by the time we gathered on the deck. During the remaining minutes before the dredge broke the surface, our leaders explained the procedure and outlined our tasks.

As the last nuggets of our coming duties were laid out, the apparatus heaved up out of the sea and was swung in over the stern. The dredge was three or four feet wide. Its entry frame was footed with a heavy rectangular plate which would drag a short distance beneath the bottom ooze and scoop all in its way backwards. A harness controlled its orientation. To the rear of the dredge's entry was a large net bag which retained whatever was captured.

## SEA STORY

Once aboard, the dredge was upended and spilled its contents in a slimy, squirming pile onto the deck. The pile instantly splashed outwards until various writhing creatures, having swiftly slid and slithered across the deck, lapped around our feet as we all recoiled backwards.

No time to ruminate about the fate of our fellow beings. We got to work. Our task was to sort the various creatures, determine what they were, and make a count of each type. Simple enough, and fascinating. After a couple of hours the job was done and the catch was shoveled overboard. We cleaned off the slime as best we could and retired, exhausted, back to the bunk room. Heaps of wet and reeking clothing littered the floor as we crawled back into our bunks. We would rinse out that stuff, and ourselves, later. This is the part of science they don't talk about when they recruit the young and innocent.

WHEN WE AWOKE IN THE MORNING we could see Santa Barbara Island. It's a high, predominantly flat-topped mountain, poking over 600 feet straight up out of the ocean, with vertical cliffsides all around. The island makes no concession to those who want to come ashore. Yes, there are a few rock platforms and cobblestone beaches which project out into the water, but these are not places to go near. The rock shelves would instantly smash a hole in the side of any boat. The few cobble beaches are inhabited by sea lions or seals, and they prefer to be left alone. Besides which, those high, steep cliffs make it problematic to get from these waterside places to the center of the island.

On the sheltered north side of the island there is one beachless cove where engineers have built access. This is a metal structure which projects from the base of a cliff out into the water. It is not a dock in any traditional sense. The landing structure is located where it is because this is the most protected, and least steep, corner of the island. Sheltered it may be, but the waves still lapped strongly against the massive pilings of this structure, propelling our boat rapidly up and down, and sideways, too. To go ashore it is necessary to grab hold of the wet, slick, rusty ladder at the 8 foot wave's peak and then let the boat fall away beneath you. And you have to bolt up the ladder once you grab it, so that you don't get crushed by the boat coming up on the next wave. Getting back off the island is even more difficult. Someone has to be there to catch you the moment you let go of the ladder so that you can fall back into the boat, not be left dangling in midair as the boat falls away.

Credit goes to our boat's pilot for making things as easy for us as possible. He spent the landing and re-boarding times juggling the throttle and rudder to keep the boat almost stationary with respect to the ladder.

The engineers had also carved a steep, staired, path which snaked up onto the island's plateau high above. The upper island is not quite flat — there is a ridge line on the south side. There is also a declivity on the southeast which slopes steeply down to a flat area edged with a cliff of modest height which rises from the sea below. At the foot of this cliff is a cobble beach inhabited by sea lions. Without going into the water this is the closest one could get to those animals.

The island's area is only a square mile so it doesn't take long to explore all that is on the land. Aside from the sea creatures this summer season hadn't much to offer in terms of sightseeing — just semi-barren grassland. The island's ranger briefed us on the wildlife here. In the spring, he said, the land is densely carpeted with the most extravagant pageant of colorful flowers. The spring, too, is the season for nesting birds. Nearly a third of the seagull population for the entire Pacific coast nests here in the late winter and spring.

Other than look after the wildlife, what does the ranger do during the year? He keeps the lighthouse running, we learned. That was not a stressing task as the lighthouse proved to be little more than a beacon on a pole. Nothing fancy. Then someone asked the taboo question, "Doesn't it get lonely, being out here all by yourself?" "You should see the wildflowers in the spring," was his curt reply. Time to move on.

THE VOYAGE TO ANACAPA, our next stop, occupied the evening and the night. An interesting social dynamic quickly developed with men and women sharing the same bunk room and changing clothes together. This situation would have been unthinkable before the '60s modified society in a fundamental way. Even more remarkable, this was under the sponsorship of the University of California, a branch of the California state government.

It was only natural that people started pairing up with evident romantic intensions. For a while there was a bit of mystery. I noticed couples lining the starboard rail towards the bow of the vessel. Pair at a time they would disappear and not return — until sometime later. The mystery was cleared up when someone started talking about the bed arrangement down below the forward hatch. Obviously there was substantial extracurricular recreation underway through most of the afternoons and evenings.

## THE TREK CONTINUES

This got me thinking about the Mayflower, a ship only a few feet longer than our own. A hundred colonists were crammed, for more than two months, into a room not much larger than the one we shared. The travelers inhabited this small room, day and night, through endless violent storms and much sickness. Our experience, in the few days we were together, showed inhibitions rapidly disappear in these circumstances. There would likely have been little the Pilgrims did not know about each other — at the most intimate level. To survive, and keep their sanity, they would have had to coalesce into a *very* close extended family, a family built on total trust. This intimacy is something most definitely not reported in school histories. But it would have served them well for the first few difficult years of colonization.

In olden days, even up to the time of the Mayflower, people believed that the western oceans swarmed with great sea monsters. Antique nautical charts warned *there be dragons here.* And indeed there were such monsters to drown unwary ships. Anacapa was one of them. From a distance the three disconnected mountains of Anacapa resemble the snaky coils of a giant sea serpent — a serpent waiting to devour hapless mariners in shipwreck, as Anacapa once did. As we drew near we could see what a monster Anacapa truly is, with its sheer cliffs and jagged rock footings — sea serpent's teeth which could crush unfortunate ships and sailors. The establishment of a proper classical lighthouse there put an end to future maritime disasters at Anacapa.

Between West Island and Middle Island there is a patch of land, an isthmus which joins the two islands together... well, sometimes. Like the mythical island of Hy Brasil, this piece of terrain appears and disappears beneath the waves, though here at the whims of the moon and sun and their tides. Our Zodiac slid far enough up the northern gravel beach that we could land dry shod. The little boat was shoved back into the sea to pick up its next load and we wandered off towards the south shore. Perched up on a nearby cliff, well above the highest tide, was the ramshackle cabin of Frenchy, the islands' longtime sole settler, now departed from this Earth. To no one's surprise, the place we had landed was named Frenchy's Cove.

Be on the lookout for moonstones, we had been advised. Indeed these mysterious gems were scattered widely along the southern gravel beach and easily found. How did they come to be? As the waves gradually erode the foot of the cliff, chunks of feldspar occasionally tumble down, shattering on the rocks below. The churning surf and densely packed plebian pebbles pound them further into small rounded and polished gemstones. Micro-laminations within give these stones their opalescent glow. The Greeks and the Romans had a different creation theory: To them, these were petrified moonbeam teardrops. Some modern eccentrics still preserve this millenniums old mythology. Anacapa moonstone jewelry can bring stiff prices among the faithful.

Despite it being summertime, the day was overcast, with a chilling wind that bit through our thin clothing. Only too glad when we were recalled back to the boat, we warmed up in the galley with a hot lunch as the boat cruised back many miles to the east. Our destination: East Island.

As at Santa Barbara Island, the landing was from the boat directly up onto a steel structure. It wasn't quite as precarious, thank heaven. Our pilot tucked the boat into a little cove, only a bit larger than our vessel, but big enough to provide some shelter from the open ocean. So the rise and fall of the boat was less, and our landing, and our later return, were not as precarious.

Forewarned by our morning's experience we were warmly dressed as we climbed the steep stairs up to the island's plateau. Before starting on the trail that circles the island we were cautioned to stay on the trail — disaster followed those who had not heeded the warning.

This area is the primary nesting place for the Western Seagull — a bird which nests on the ground. (This made sense since there were no trees or bushes in the area.) The trail winds through the nesting area, but in such a way as to not intrude on the actual nesting sites.

Some years previously a party of nature lovers, sponsored by a major environmental society, had been given the same warning. But they did not listen. The springtime nesting birds were so darned attractive! Those people, mostly middle-aged ladies we were told, could not resist the temptation to walk up to the nests and start photographing the birds. Naturally the nesting seagulls took fright and flight. The whole colony, more than ten thousand birds, spiraled up into the air. Hovering nearby was a hungry pack of skuas. The gulls now in the air, the skuas dove in. They did not land. Instead the skuas flew low over the nests and pecked the eggs as they passed by. It was over in minutes. The entire breeding colony had been wiped out for the year. Feasting followed — on the eggs of two thirds of all the young seagulls for the entire West Coast. Fortunately, the much smaller colony on Santa Barbara Island still existed, and made up some of the loss.

However, the prominent environmental organization was banned from Anacapa Island for years. Also, for a long time thereafter no tourist groups, or casual visitors, were allowed on the island. We, being from the University of California, were the privileged exception.

I am happy to report that in the years since, the colony had been rebuilt and the nesting sites were still occupied, even though that season's eggs had hatched. The young birds had grown quickly and were intermingled with their parents. The enormous mass of birds was raucous!

## Santa Cruz Island

IT IS NOT FAR FROM ANACAPA to Santa Cruz Island, the next land mass on our route. By late afternoon we had temporarily anchored near some sea caves on the rocky northern shore. It was time for a treat, but that would take a while to materialize. Earlier we had had an interesting experience along the island's sea wall. The northern edge of the island mostly consisted of basalt cliffs — the long-ago product of marine volcanoes. Weak spots in these rock walls progressively failed in response to the constant pounding of the sediment-laden ocean waves. Gradually, at these varied locations, the rock eroded away. The erosion was not uniform, however. The resulting sea caves were funnels tunneling deep into the cliffs. On a calm water day it was safe to proceed a considerable distance into these tunnels. On a day like this it was anything but safe.

The earlier promise to take volunteers on zodiac rides into the sea caves was broken. Not today. We watched from afar as a large wave rolled its way into a cave. It did not break because the floor of the cave was too deep. But the cave *was* a funnel, so the wave was squeezed inwards from the sides as it plowed on. The wave rose in response — higher and higher as it plunged further into the narrowing darkness, the air inside hissing out in anger. The wave reached the ceiling, sealing off the air inside. Still the wave advanced and then... exploded! The air in the cave had been compressed until it was more powerful than the incoming avalanche of sea water. A massive jet of wet, white foam rocketed out from the opening to drench us all, even though we were well away from the mouth of the cave. Our instructors burst into laughter.

Eventually we dropped anchor along a quiet stretch of sea cliff. It was the perfect spot for what was to follow. Soon a couple of black clad crewmen appeared on deck, weight belts sagging, knives and other implements strapped to their legs and arms. They went through the usual rigmarole of checking out their scuba gear, testing their regulators and tanks and clearing their mouthpieces before strapping on their fins and helping each other mount the heavy apparatus on their wetsuit-clad shoulders. Then, falling backwards, over the side they went, and down. We soon lost sight of them as they disappeared deeper into the cold, crystalline water. Only bubbles marked their descent.

On the aft deck a strange apparatus had been mounted — a large steel box secured on standoffs. The box was lined with bricks and filled with charcoal briquettes. We were about to have a barbeque — on board a rocking and rolling boat. Actually, there was nothing new in this. Shipwrecks tell us that fireboxes like this have been the standard way of cooking on vessels for thousands of years, back all the way deep into the Bronze Age.

The divers reappeared and a net bag filled with some sort of bounty was hauled up onto the deck, followed by dive tanks and fins and ultimately the divers themselves. The treasure that the divers had brought up from the bottom of the sea was abalone — *large* abalone. Because of overfishing along the California coast such foot wide shells were almost unheard of. But that was then and this "secret" spot north of the cliffs of Santa Cruz Island was, in that time, still untouched, and legal. At the time of this writing it's illegal to harvest abalone, due to a die-off from warmer waters as well as poaching, though there are efforts being pursued to bring back abalone in the wild.

Dinner was still more than an hour off. First, the shells had to be shucked, the meat prepared, and the sliced off steaks tenderized and marinated. This took time, with the cook working on a wood-topped bench off to one side. Finally, kerosene was poured over the charcoal bricks and, with a whoosh, the firebox was set alight. Soon the coals were visibly glowing red in the dim afternoon light, the bright sun having been shaded by the nearby cliff.

Dinner was served — grilled steaks and fresh abalone — together with fried vegetables, served on skewers. Delicious! A Top of the Mark dinner. On occasion they treated us exceptionally well during this academic excursion.

After that great meal, and the cleanup, we raised anchor and coasted along the extensive length of Santa Cruz Island. The sun had already slipped into the sea by the time we dropped anchor in the sheltered area between Santa Cruz and Santa Rosa islands. Tomorrow we would be off to visit San Miguel.

The following day fled past and we found ourselves still between Santa Cruz and Santa Rosa. It was the last night, the one on which we were anchored among the Russian fishing boats. We talked about the day we had just spent on San

Miguel Island, and our many other adventures. But most of all, we talked about the Catalina scream. It was amusing, now. Not at all at the time.

It happened the second night when we were camped on Catalina. I remember the hard ground becoming soft as I drifted off to sleep, deep, dreamless sleep. Suddenly, an agonizing shriek jerked me awake. Hot adrenalin jetted through me as the scream went on and on, echoing from nearby Black Jack Mountain and even, faintly, from more distant Mount Orizaba. Flashlights popped on from all over the camp site as the more nimble rushed towards the unearthly cry. Then there was silence, followed by the rumble of voices, and even some laughter. Someone called out to the campers that all was well, no harm done. Before the adrenalin rush had fully subsided, so that I could drift back to sleep, an instructor walked by and reported that it had merely been a bad dream. Nothing to worry about, we will talk about it in the morning.

Bad dream indeed! Imagine you are deep asleep. You begin dreaming. Nothing special at first. Then a large formless shape approaches. You try and escape, but the shape traps you, pins you down. The shape's great mouth, filled with wicked teeth, opens and the shape's head comes down upon you.

You wake up with a start. You are trapped, and the large formless black shape *is* eating you alive. You feel its tongue ripping across your face, its slime floods into your choking lungs. You are pinned in your sleeping bag. The giant creature straddles you as it starts sucking the life from you. You are fully awake, intensely awake, and screaming!

This is a true story. All was as I have described. The explanation is simple — but still scary. The victim did have such a dream. He did awake to find a giant monster straddling him with a huge tongue, from an enormous mouth full of teeth, slathering across his face.

The monster was a buffalo. It had carefully walked up along the length of his sleeping bag and proceeded to lick him awake. Startled by the scream the buffalo delicately disengaged, again careful not to step on its victim, and wandered off. It did not return that night.

That last night aboard, the one between sheltering islands, among the boats of the Russian fishing fleet, we laughed again about the scream. In the morning we sailed for home. Dolphins played in our bow waves. Great whales accompanied us and wished us bon voyage.

We had had an adventure.

*A herd of Buffalo innocently grazing on Catalina Island*

*Chester at sunset after a day of surfing at the beach*

# THE PERFECT WAVE

## PART 1

THE SIREN SONG OF SURFERS *is the search for The Perfect Wave. This gift is given to few and seldom given more than once.*

*Although a surfer may dream of The Perfect Wave, the dream is only a fantasy of what it might be like. One does not know the reality until the ride of a lifetime has actually been experienced. So it was with me.*

I WAS NOT A GREAT SURFER. But with years of experience I became deeply knowledgeable about the many moods of the sea, and its waves, and the individual personalities of my favorite surfing spots. Most of all, I had fun — great loads of fun.

At first I had a big board for stand up surfing but after lugging that heavy thing around I was often too exhausted to really play in the waves. The solution was simple: I switched to an experimental paipo board. This precursor of the modern shortboard was about four feet long with a beautiful yellow glass surface and a crescent skeg. It was light and small enough to fit in the back seat of my VW bug.

Even better, its small floatation made it easy to push under the breaking waves of the bone yard. The bone yard is a seriously scary place when the waves are really big. You can drown in the bone yard. Let me explain.

It was back in the late summer of 1958. I was nearing the end of high school, but still had a year to go. A family excursion took the long, pre-freeway drive down from La Canada to Laguna's Crescent Cove, a mile or so up the coast from the main beach. We arrived on a brilliantly sunny day and walked down the steep hill to the beach and the blue water beyond. The surf was excellent for body surfing — about five to six feet with clean, only slightly wind-rippled waves. But this day the lifeguard station had two red flags flying, not just the single red flag which normally flew on big surf days.

We spread out our beach blanket. Cautioned by the two red flags, I did not go in the water immediately. I waited and watched for about half an hour. Sometimes the waves increased to eight feet or so, and even an occasional ten footer would sweep in, but these were within my experience and I was comfortable with the prospect. So, I grabbed my fins and entered the surf.

For an hour I rode the waves. With fins on my feet they were easy to catch. Being large waves, they gave a good, long ride. These waves broke in deep water so there was little hazard in tucking under and flipping out of a wave if it suddenly elevatored with the beach's backwash. I was having a very good time.

Every once in a while a big set would roll in. From the water a big set materialized as a dark blue line against the horizon. "Outside," someone would yell to alert the rest of us, for these big waves would break outside the usual lineup for the common waves. The gang of experienced body surfers, which always forms on a good surf day, would then race out to

# THE PERFECT WAVE

sea to get to the new surf line before the first waves of the big set arrived. This was always a good moment because bigger waves gave better rides.

Except, except, not this day.

I had just worked my way back out to the surf line after a really good ride on a ten footer when, far out to sea, a *black* line stretched across the horizon. I had never seen anything like it. These must be really big waves. I looked around. Half my companions decided to head out to meet the new set, the other half, the wise half, headed for shore. I was one of the fools.

I had no intention of riding these giant waves, I just wanted to get beyond the surf line so that I would not get caught in the bone yard. The guys heading for shore were sure to be badly pummeled by the deep, powerfully turbulent froth of the broken waves. I didn't want that.

This was not a good situation. I raced out to meet the incoming swells. The first wave loomed larger and larger as its foot caught the feeling of the deep water bottom sands, far below. I put on an extra burst of speed. This was a close race now.

I didn't make it. I was only a few yards short of the towering swell when the giant wave started to break. I dove under so that I would not have to face the force of the crashing waterfall. Diving deep is the standard tactic when one is about to be caught in the bone yard. Usually the elliptical motion of the wave's water will suck you right through the wave so that you pop up, safe and sound, on the wave's back side.

This day the tactic did not work. This wave was far too big for that. Even deep under water I found myself still on the front side of the wave. Instead of traveling down and through, I felt myself being accelerated up, high up, and backwards.

## THE TREK CONTINUES

I went over the falls, tumbling head over heels in the white froth of the collapsing giant.

My fins were stripped from my feet and I was pushed down, down, *down* to the bottom of the sea. Twenty feet, thirty feet, maybe more, I'm not sure how far down it was. But it was *deep*. The intense pressure on my ears told me that, so did the dimming of the light.

I cleared my ears and tried to push off the bottom. I could not rise — the wave held me pinned there. I did several push-ups off the bottom, trying to break free and rise back to the surface. Finally, I got smart and just curled up into a ball, relaxing to conserve oxygen. The wave still held me pinned to the bottom sands.

In those days I played woodwinds as a semi-pro musician. This had developed my lungs to the extent that I could hold my breath, and still be active, for up to four minutes. A minute on the bottom was not that much of a problem. But I did not reckon what was to come.

Eventually even the biggest wave passes and the pressure releases. I pushed off the bottom and headed up to the light. I reached it, but to my horror, I could not break the surface. I was trapped in the foam of the broken wave, still several feet under, and you can't float up through foam. I needed to reach the surface — I couldn't breathe.

Another long minute passed before the thick layer of air bubbles migrated to the surface and I could at last reach life-giving air. My head rose fully into daylight. I gasped and drew in a single deep breath just as the fall from the next giant wave crashed down on me. Back to the bottom I was driven by the powerful downwash. This time I just curled up into a little ball and let the wave's power toss me about like a toy.

Again I reached the surface after minutes beneath the sea. Again I could gasp just one quick breath before the next wave drove me deep. Five times, altogether, I was driven to the bottom before King Neptune relented and decided to throw me back to the living.

Exhausted now, exhausted almost beyond life itself, I slowly dog paddled my way back towards the distant shore, my arms too heavy to lift out of the water. The new surf line seemed infinitely far away, but I eventually reached it. Its breaking waves now became my friend. I deliberately let a wave crash over me. Its turbulent wash pushed me still closer to shore. Again, a breaker picked me up and swept me closer to the strand. Finally I washed up on the sloping beach and just lay there, still half stunned, my fingers clawed into the wet sand to anchor me.

I lay there drinking in warmth from the sand and sun, while life's energy slowly poured back into me. Two bronzed, blond-hairy legs walked towards me and a red suited lifeguard stooped, briefly, to check me out. No questions were asked and the legs walked on. Annoyed at this lack of concern, I flopped my head to the other side. There, stretched out in a line, was a row of bodies, each clawed identically into the life-giving Earth.

Alive at last, I grasped the humor in the situation.

## PART 2

**NEPTUNE IS A HARD TEACHER.** Far up the coast, at the Huntington Beach Pier, these enormous waves I had endured that day at Crescent Cove swept over the pier and completely demolished the tavern perched at the end. I have seen twenty foot waves break *under* the pier. For waves to break over they must be near thirty feet. My waves, known for many years after as the "All Time," were true giants. But I learned Neptune's lessons well, lessons that would save my life again and again, in the ocean and in violent rivers.

A few years later, graduate studies brought me to the newly opened University of California campus in Irvine. I took up residence in nearby Costa Mesa. This put me a short drive from the pier at Huntington Beach, which soon became my home in the water.

Most days, year around, the surf near the pier was five or six feet. From late summer until late autumn a big southern swell would arrive every few days and the surfing became terrific. The pier, though, was a problem, until you became used to it. Then it became a friend. Massive concrete columns rose from the water to block the surfer's path. These columns were encrusted with razor sharp mussels and barnacles. Even a slight brush with a piling could slice you open to the bone. And there was always the possibility of a direct collision as the wave swept you remorselessly in. I was certainly intimidated in the beginning, and later I had my share of accidents. But I watched, and watched some more the experienced surfers as they wove their way effortlessly through the forest of columns. I finally figured out what they were doing.

Eventually I had to try it myself. That day I was on the south side of the pier. The waves were good sized sets coming from the southeast (the coast runs almost east-west here), marching in regular lines up the coast. These waves were entirely predictable so there was no better time to try out this insane stunt.

## THE PERFECT WAVE

I lined up and took off left, well up on the face of the wave. The wave swept me rapidly towards the fast-growing line of pilings. As I approached I discovered one of the unexpected secrets of Huntington Pier. The pilings are surprisingly far apart — much farther apart than they appeared from a distance. I had no trouble maneuvering to clear the nearest one. Then I discovered a second secret. The surf did not break around the pilings. The wave lost its energy as it approached the deep scoured bottom near the pilings and let me down gently into relatively placid water under the shade of the decking. Then came the third secret, and the greatest surprise of all. There was a strong rip current here and the current ran straight out to sea.

Hallelujah! The pier was like a ski lift. Slide under the pier and its current would whisk you right back out to the surf line. If you surfed the south side, the pier current, combining with the regular south going shore current, quickly propelled you back to your starting point on a nice merry-go-round — no fuss, no muss, no bother — and with little energy expended. If you surfed the north side you simply let the pier's outwash take you a bit further out, beyond the shore current, and you then could quickly paddle back to the north side lineup. The more I surfed through the pier the better I liked that place.

I liked surfing the south side, with its clean, predictable lefts. I liked surfing the north side even better. The north side was less predictable because the pier broke up the long waves from the south and reformed them into shorter sections. But north side had the great advantage that somewhat smaller waves could also impinge from the southwest at the same time as the large southeast swells. On days when these waves from two directions collided they created a piled up peak with good rights and lefts, both — your choice. If you chose right the wave would take you to the pier and the ski lift there. If you chose left the wave would take you to the edge of the north side rip tide and another merry-go-round which would circle you right back to the north side line up. I can't think of a better place to spend long, leisurely days in the late summer and into the fall.

The summer passed and the new school term opened. I was working late at night doing research using the University's main frame. This left my days pretty much free to surf, although with some penalty in lost sleep. Surfing in that season was near perfection. School and work cleared the waves of competitors. The air was warm, as it usually is in California's autumn. The big south swell was particularly good that year. The winds were mild so the waves remained clean, not windblown, until late in the day. One could hardly ask for more. This was pure Paradise for any true water baby.

One sunny morning, much too early for I had been up most of the night, King Neptune whispered in my ear: "Wake up! Get out of bed, lazy bones. I've got something special for you — a gift. *Get up!*" Wearily I dragged myself out of bed, rustled up some breakfast and prepared to mosey up the road to Huntington Pier.

I drove up the back way to Huntington Beach. Nearing the pier I found a parking place only a block away. This is the closest I had ever been able to park. I walked the short distance to the pier, catching glimpses of the waves across Pacific Coast Highway. From this distance the waves seemed immaculate. I crossed the highway and worked my way down the steep south side stairs, across the broad sand desert of the beach, and into the water. Being savvy, I used the ski lift under the pier to take me out past the shore break, then much further out, past the outside breaking line. I caught the beach current and drifted down to the south side lineup. Unusually, I was alone with the waves.

The ocean had warmth with a late season current bringing tropical water up from Mexico. The hot sun on my bare back felt good in contrast to the coolness of the ocean. A regular, wonderfully angled, south swell of good size was marching up the coast with machine like precision, wave after wave. There was the slightest off-shore breeze, just enough to brush the waves up to nice peaks, but not so strong as to prematurely blow the tops off the breaking waves. With an off-shore breeze the water was glass smooth. All alone with these beautiful waves, I started playing.

Ride after ride I skimmed across the faces of these wonders. With the large size of these waves the shore current was stronger than usual so I soon lost touch with the pier and its merry-go-round. It didn't matter, I let myself drift south. This is what Neptune wanted for me, it was part of his surprise. The waves were so perfectly formed and so nicely brushed up by the gentle breeze that it was no trouble punching through them and paddling back to the surf line after a fine ride. Gradually I drifted down the coast until the distant pier formed only a small segment of the horizon.

I was now more than half a mile, maybe even a mile, south of the pier. I rose up over a large wave and surveyed the prospect out to sea. There, on the horizon, was a wave bigger than any I had yet seen. Quickly I started paddling out, my fins churning the water, my old board planing over the surface. I could see the great swell starting to loom. Unlike my earlier experience at Crescent Cove, this time I was in excellent position. This was a wave I could ride.

How big was the wave? It wasn't the giant that had smashed me to the bottom of the sea and had demolished the outer portion of the Huntington pier. This was much smaller, but still big enough. Perhaps it was fifteen feet. Yes, about fifteen feet. What is more, it was perfect. Long, very long, I could see it breaking a mile down the shore from me. Angled, perfectly angled, the breaking front propagated smoothly up the coast without interruption or snags.

I waited for the wave front to reach me. The wave started lifting. I turned and stroked for all I was worth. I started sliding down the incline and quickly picked up speed. Fast planing, now, on the shoulder a few feet ahead of the break, I pulled myself to the front of my board and settled in for the ride. A slight adjustment in direction and the wave lifted me higher, and yet higher, as my speed increased still more. I cut slightly right and settled into the groove, maybe three or four feet from the wave's crest. There I stayed, firmly locked in, flying faster across the face of that steep wave than I ever had in my life.

Was there going to be no end to this ride? I sliced through the water, on and on and on. The distant pier started to grow as I glided effortlessly in its direction. People on the pier caught sight of me and called out to their companions. I could see figures running from all parts of the pier to watch me. I did nothing — just lay in the groove, making minor adjustments. The wave did all the work.

A quarter of a mile flowed past. A half a mile. The pier was growing ever larger. All on the pier were jumping up and down, waving their arms. They were enjoying my ride almost as much as me.

The wave seemed endless.

But all things must end. All too soon I was approaching close to the pier. I began planning the tactics of my passage between the pilings. I now could hear the screams of delight and encouragement from the structure's deck, not far above.

The wave started to feel the increasing depth around the pilings and began its death dive to oblivion. Still, it had force enough in its dying struggle to propel me like a rocket through the pier and out to the north side.

The crowd ran across the pier to watch me emerge. I turned and looked up. We saluted each other, all of us amazed at what had just happened.

As he had promised, that day King Neptune gave me his gift, the gift of *The Perfect Wave*. There were no cameras present that day, no film crew to record the extraordinary event. There are now only the ancient memories of the few who were there that day, and these words to tell the tale.

# EVERY DAY IS AN ADVENTURE

THERE ARE DIFFERENT CATEGORIES OF ADVENTURE, I find. Some are quite deliberate, a few of which we just swam through in the Water Babies section. There are also what I think of as everyday adventures. The kind that we do not seek out, but that just happen to or around us. These are the kinds of adventures I'd like to share with you here.

There are stories from my youth. There was nothing particularly unusual about my childhood. Most of my generation share a common experience of school and everyday living. The only difference from today is that kids then had more freedom — *much* more freedom. I now know that daily grown-up life is mostly routine. But through the now-long years some things are sufficiently quirky to stick in memory and become worth the telling.

Then, too, there are the unexpected hazards that one somehow survives. I have had my share of those. And readers tell me some of them are quite capable of raising anyone's hair.

AS WITH MANY PEOPLE OF MY GENERATION, most of my early life was spent in school — ultimately toiling away through long years at the university. I will soon tell a tale here about graduate school and why I ultimately walked away from it. As I mention in the story Failures: Part 3, leaving the University when I did turned out to be the second best decision I ever made (wedding Sarah was number one). Abandoning my Doctorate saved me from being smothered by the colorless, stifling miasma of academia, where you are forced by specialization and the academic grind into doing the same thing over and over again. I am most definitely not suited for that world. Now I found myself free to develop in fundamentally new ways. A Doctorate is nice to have as a door opener, but there are other ways to open doors, I soon discovered.

It later occurred to me that I became fully an adult in the moment when I decided to leave school. No longer was I preparing for life. Now I was to be immersed *in* life – big time! I had spent two years working as an engineer between undergraduate and graduate school. Now I was back in that familiar and comfortable real world where there was almost always good humor and I could laugh with my friends and colleagues as we beat our heads against seemingly impossible real-world problems.

The years before that momentous decision had its charms, though. Despite what felt like failures at the time.

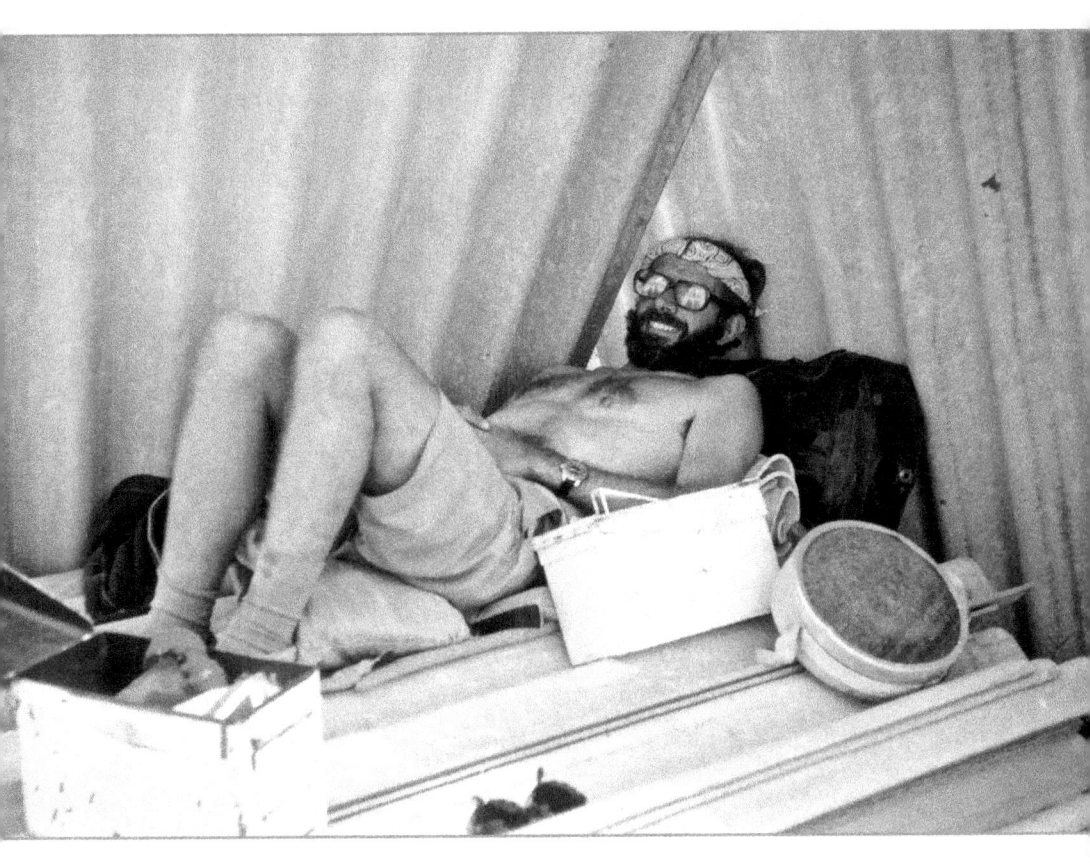

*Chester getting foot treated after scorpion got him*

*A U.S. Air Force Fairchild C-119B-10-FA Flying Boxcar of the 314th Troop Carrier Group in 1952. The type of plane on which I was a passenger in this story. This aircraft was later converted to a C-119C in 1955-56. The 314th TCG served in Japan during the Korean War, participating in two major airborne operations, at Sunchon in October 1950 and at Munsan-ni in March 1951. It later transported supplies to Korea and evacuated prisoners of war.*

*Original caption: "The 314th Troop Carrier Group C-119 Flying Boxcars do not start out for the 'mountain' unless weather reports are good. They must be able to see the tiny drop zone on the peak before they can drop. But weather is so unpredictable in the high mountains, that often when the planes arrive, the entire area is 'socked in' with heavy clouds. In the radio contact with the "men on the mountains," the pilots circle hoping for a break in the clouds, or sometimes, to dive under the clouds and drop on the lower slopes. On several occasions, the planes have had to return to Japan as many as three times without dropping. But 314th Troop Carrier Group pilots are persistent, and eventually win through to drop successfully. 1952 (U.S. Air Force photo)*

# WHITE KNUCKLE FLIGHT

A WEEK BEFORE IT HAD RAINED, a cold, cold rain. Then it froze. All week long the temperature hovered near zero. Ice covered all. The old timers at the University of California's Berkeley campus had seen nothing like it before.

The Berkeley campus lies on the side of a hill, its upper parts being quite steep. Some thoughtful person had rigged up a long rope to assist people on the slippery walk down the incline past the crumbling old red brick chemistry labs, past Le Conte Hall where physics was taught, and down to the now vanished elegant rotunda of Bacon Hall and an area of level ground.

One Saturday when I was a freshman, it was time for a long excursion by air down to the factory in San Diego where the Atlas missiles were being fabricated. These excursions were one of the perks of the Air Force ROTC in which I was enrolled.

I rose very early that day, grabbed a quick box breakfast that I had arranged to have waiting for me at the Bowles Hall dining commons, and slip-slided my way down the long hill, half hanging onto the rope and half ice skating with my spit polished Air Force shoes to where the bus was waiting. My blue gabardine Air Force uniform was much too thin for this bitter cold.

The bus headed to Hamilton Air Force Base some distance north of San Francisco. It was a long ride. We finally arrived and

## THE TREK CONTINUES

dashed from the warmth of the bus to the warmth of the Air National Guard ready room. After a brief check-in we began the long walk through the bitter cold, over the ice covered tarmac, and out to where our aircraft was waiting.

Near the aircraft was a pile of parachutes. We stood shivering in formation while the flight sergeant instructed us in the proper use of these parachutes and aircraft exit procedures. He assured us that this briefing was just a necessary precaution in case of engine failure, which was very unlikely. But he also told us that should we be required to jump we should get out in a hurry because the plane glided just about like a brick. Finally, after interminable instruction, we each grabbed a parachute and were hustled out of the cold breeze and into the refrigerated interior of the waiting transport.

The hulking plane we now entered was a Fairchild C-119 Flying Boxcar. It did look a bit like a boxcar, a silver boxcar, with wings. Inside, the cargo compartment walls were lined with canvas benches and a bit of padding. We strapped on our parachutes and sat ourselves helter-skelter along the sides of the cargo space. The entry doors, set into the clamshell doors in the rear of the cargo pod, were hauled shut and locked. The engines coughed into life. At last, after a few minutes of engine warmup, we started to get some heat back into our bones.

Takeoff was routine. We slowly climbed to altitude, passing briefly through the high cloud cover into the sunny blue, and headed south. This plane was not the swiftest thing in the air. It struggled to cruise at a couple hundred miles per hour. We settled in for what promised to be a long flight, at least three hours to San Diego. As the plane droned on, we unbuckled our parachutes and started to move around the cargo space to look out the tiny porthole windows on both sides. San Francisco was attractive, as always, but soon the repetitive scenery lost some of its charm. The noise precluded easy conversation, so most of us simply sat and waited, bored, with nothing to do and little to see. The flight sergeant climbed down from the cockpit and rustled up some coffee for the crew. Some of the cadets were given coffee as well.

An hour passed. Nothing to see, nothing to do. We were far out over Monterey Bay now, taking a short cut over the inky arctic blue waters of the Pacific Ocean. The copilot climbed down the ladder from the cockpit and started rummaging around in the little storeroom below it. As he backed out of the room he was tightening the straps on the parachute he had just donned. He climbed back up. This struck us cadets as not a good sign. Those of us who had been wandering around the cargo area started edging back towards our seats and our waiting parachutes. Then the flight sergeant climbed down and ducked into the little room. He came out wearing a parachute with a chute for the pilot slung over his shoulder. Up he went and disappeared through the cockpit's door.

By this time all of us were busy buckling on our parachutes and cinching up the straps to the prescribed tightness. The plane did a shallow wingover and headed back north. A red light winked on.

Now, this particular plane had a three light warning system for bailing out. The first light was red. This light said to get into the parachute harness and tighten up. The second light was also red. This second light said to stand up in two lines and hook up to one of the overhead metal wires strung down the length of the fuselage. The rear doors would be opened on the second red. We had been instructed to crowd up to the open doors and be prepared to jump. The third light was green. When the green light came on you jumped, no hesitation, no questions asked. Get out now! Remember, this plane is a flying brick.

The only good thing about a C-119 is that it was specifically designed to deliver paratroopers. The tail empennage would be well above our heads should we be required to step out of the rear of the plane. Small consolation.

Sure enough, the second red light came on. The flight sergeant hustled back and swung open the twin rear doors, pulling them inwards. We stood and hooked up to the line and edged forward. I was positioned second from my door. Through the aperture I could look down to the winter-cold water below. We were still dozens of miles from the nearest land.

The howling wind from the open door masked the previously unbearable drone from the engines. Bitter cold swirled around and sucked the warmth from our lightly clad bodies. As I stood there I imagined what was about to happen:

The green light would go on. I would be pushed out the door by my friends behind me —no choice in the matter. As I left the plane the free fall would tumble my stomach into butterflies. There would be a short jerk as the static cord ripped open my parachute pack. A slight drag for a couple of seconds, then a major shock as the full canopy cracked open. My dress shoes would instantly be flung from my feet and the freezing air would start to numb me. I would grow dazed with the cold as I glided down to the ocean below.

Then, into the water. Further shock as the frigid water swirled around me, soaking through my clothes and parachute harness and dragging me toward the bottom of the sea. With great luck I might be able to break free of my parachute, shed my jacket and trousers and rise to the surface. But I had no flotation. Maybe a life raft would be nearby, and maybe I would be able to swim to it. I knew that I would not have the strength to pull myself into it if, by a miracle, I did manage to reach it. Within minutes I would be losing consciousness from hypothermia. Before rescue could reach me I likely would be dead. Pleasant thoughts as I gazed through the open door down to my doom.

The plane droned on. We neared the shore. Perhaps I could survive a dry landing, though without boots, or even shoes, my ankles would be smashed, and probably much else, besides. I would be badly hurt but I might live. Some hope started creeping back into me.

We stood there, at the open doors, for what seemed an eternity, though it probably was only a few minutes. While we were waiting the flight sergeant went up to the cockpit to check things out. He came back after a bit and closed the rear doors. The whistling wind grew quiet and he was able to shout that we should return to our benches and sit, but to keep our parachutes on and tight. No one argued.

Now came the long ordeal. Now we had time to think. What exactly had happened? What had gone wrong? How much danger were we in? No one spoke. All had pulled within themselves to review what had just happened.

The flight sergeant returned to the cargo bay and shouted that we were on our way back to Hamilton, that there had been an engine problem but now the pilot thought we could make it there okay. Just in case, though, stand by to bail out if need be. This time, get out as fast as you can. Not comforting words. The plane droned on.

Now, it seems to be the nature of men in dangerous circumstances that a particular form of humor surfaces. Call it gallows humor. That instinct came to the fore and various tight-lipped jokes were passed around, shouted over the roar of the engines. Perhaps if all were normal the jokes would not be very witty, but there, in that moment, these few words were falling down funny. We waited some more, the tension rising as we neared the airfield. The jokes grew wilder.

The aircraft started a wide, shallow bank to the right. Very wide, very shallow, something was indeed very wrong. Eventually the wings leveled out. The flight sergeant dropped down from the cockpit to brief us. Shouting above the engines he informed us that, since we had left, Hamilton had become socked in. Visibility was zero right down to the ground. We were going to have to make a GCA (i.e. radar) landing. We should know that this was dangerous, and we might not make it. We only had one chance at landing. If we didn't touch down properly we would have to ditch in San Francisco Bay, beyond the end of the runway. The location of the life rafts were pointed out to us, along with instructions on how to deploy them. The sergeant told us to remove our parachutes — it was not safe to wear them now — and stow them under the benches, tied securely in place. He wished us luck, and returned to the cockpit.

Again we sat, buckled in tightly, as the aircraft slowly drifted down from the heavenly blue towards the blanket of white below. I was located beside one of the few windows so I had a good view outside. The first tendrils of cloud fled by as we dropped through the surface of the white sea of clouds. These clouds were thick. Within moments I could barely make out the wing and engine boom. The light grew dim, gray and gloomy. It grew darker as we plunged further into the depths.

The jokes petered out, suppressed by the growing gloom of our descent.

It grew as dark as deep twilight, as if ultimate night were swallowing all the light. There were no stars in this fearful night. Nothing could be seen outside, not even the nearby structure of the aircraft. Outside there was only deep, dark gray void. It grew even darker.

Suddenly, without warning, we dropped through the bottom of the cloud bank. We were barely above the runway, maybe fifty feet, maybe less. We were perfectly aligned with its center. Stretching away, right across to the edge of the runway was a line of fire trucks, ambulances and other emergency vehicles, pacing us with military precision as we descended the last few feet to the runway. I glanced out the window across the cargo bay from me. Fire trucks were there, too.

We touched down. An audible sigh of relief rose above the noise of the engines as we all simultaneously let out our long-held breath.

What had happened? Later we found out. Over Monterey Bay the oil pressure on both engines had suddenly dropped to zero. There was no oil left in the engines, except for a remaining thin film which, by some true miracle, kept the engines going. It really was touch and go the whole way back. The engines should have seized up immediately, and could have at any moment thereafter. With its poor glide ratio the old Flying Boxcar would have simply fallen out of the sky, taking us with it.

# MY HERO

**A** CROWD OF EXCITED BOYS CLUSTERED around a handsome, athletic man. I walked over to see what the fuss was all about. What happened next was an unforgettable surprise.

It was nineteen-fifty. I was eight years old. Warm, prosperous, dynamic Southern California had welcomed us refugees fleeing from the cold, damp, windy misery of the San Francisco Peninsula. We settled, for a while, in an apartment in Glendale.

Life was freer in those days and in that place. No one worried much about the safety of youngsters. It was no big deal for kids to wander miles from home without parental concern. Indeed, each school day I walked, by myself, the mile to my third grade classroom, crossing three very busy boulevards along the way. As I say, it was no big deal.

One Saturday afternoon I trekked up the road to Verdugo Park. The place was jumping with activity. A baseball game was in progress at Stengel Field. Casey Stengel, the illustrious manager of the New York Yankees, lived, in the off season, in Glendale. It was natural that the newly constructed municipal baseball stadium be named for its most famous resident.

New to the area, I was only beginning to learn the secrets that all the local kids knew. So I followed some boys along the fence that was guarding the first base line. Sure enough, there was a small hole in the fence. One by one each of us crawled through the barrier and worked our way back to the stands.

Seated now, I asked someone what was going on. "It's an exhibition game," was the response. That was all the explanation I was going to get. I was left to figure out for myself what was happening. This was my first real baseball game. I had seen bits of the game in newsreels but the rules and the subtleties of the sport were beyond me. I tried to puzzle out the activities playing out before me. Occasionally a roar went up from the crowd. The significance of what had just happened was still a mystery.

I had come in late to the proceedings so it wasn't too long before the players retired to the locker room and the crowd began to disburse. I left the stadium the way I had entered and wandered back towards the parking lot.

A crowd of excited boys clustered around one of the exiting ball players. I walked over to see what the fuss was all about. Not knowing anyone, I stood aside from the agitated group. The kids in the circle were a bit taller than I so I couldn't see what was going on. But I could hear several of the boys demanding to carry something. The tallest of them, perhaps a bit older than the rest, led the pestering.

Not responding to the demands, the man in the center of the crowd gazed over the heads surrounding him, looking for something. Suddenly he pushed through the mob and walked up to me. He lifted up a duffle bag with a central handle and shoved it into my belly. "Here kid, you carry this," he said, looking down at me. I grabbed the bag and nearly dropped it. It was heavy!

## THE TREK CONTINUES

We walked north towards the far end of the parking lot. A comet tail of the kids who had been in the cluster trailed behind us, the tallest boy leading them. The man walking beside me, knowing I was struggling with the burden, kindly eased his pace so I could keep up. The comet followed some distance behind us. *This* man they did not want to provoke.

The car awaiting us was something I had never seen before, or even dreamed existed. It was smaller than other cars, low slung, streamlined, its body a highly polished bright yellow and chrome, and obviously very expensive. The man took the duffle bag from me and slung it, as if it were light as a feather, behind the front seat of the two-seat convertible. This done, he turned to me and with a compliment and thanks, handed me a fresh minted silver dollar. In my hand this beauty, with its crisp sculptures, gleamed in the brilliant sunshine. It was a handsome reward for my struggles. Hopping into the front seat without even opening the door the man started the sports car and backed it out of its slot. With a flourish he drove around the corner and vanished.

"Who was that?" I asked one of the kids.

The big guy barged forward and took over, his voice raised.

"Don't you know who that was?" he almost yelled at me. "Are you some kind of dummy? Are you stupid? *I* should have carried the bag, not you."

I said nothing. I had no idea what he was so upset about.

"That," my challenger said, "that is the greatest man I know. He's my hero. *I* should have carried his bag."

I didn't know how to respond. I still didn't know who the man was.

"Who was he?" I asked again.

The big kid puffed himself up.

"That, you dummy, was Errol Flynn. He's the greatest. He is *my* hero!"

I had never heard of Errol Flynn but it didn't matter. *I* had the silver dollar.

# DIVERSION

*Firecrackers!* Tiny little ladyfingers. Regular old firecrackers. Much larger salutes of various sizes. *Cherry bombs!* Those small, green-stemmed, shiny red spheres of pure explosive power.

There were a lot of things you could do with these various small explosives. Appropriately attached to a Popsicle stick propeller, with their ends sliced open to make small rockets, a couple of firecrackers made a fine pinwheel. Or, one could carefully slit a cracker to make a whistler. The gun powder was useful, as well. In high school we specialized in fabricating ICBM's — Intra Campus Ballistic Missiles. All that was needed was a short length of soda straw, the attachment of a large size matchstick for stabilization (the match may, or may not have been burnt), some scotch tape reinforcement and, of course, the gunpowder from a couple of firecrackers. What a delight to watch those miniature rockets soar high over a shielding campus building to come raining down, seemingly out of nowhere, onto the campus quadrangle, still trailing a stream of smoke from the burnt out gunpowder. The faculty was fit to be tied. They never did figure out the source of these tiny terrorist missiles.

My favorite, though, was the coffee can rocket. A neighbor kid taught me that one. To make one you needed a half height coffee can with its lid completely removed. You also needed a soup can, with its lid completely removed as well. You punch a small hole in the bottom of the soup can and insert a firecracker most of the way through the hole in such a way that the fuse is on the outside of the can. The spurs of the punched hole hold the firecracker in place. Real artists could lodge the firecracker all the way inside the soup can to get the maximum effect. Finally, the coffee can would be half filled with water and the soup can deposited upside down in the coffee can. All was ready for launch. Light the fuse and back well away. BANG! And the soup can would rocket upwards almost out of sight — a perfect demonstration of rocket reaction forces, and a good lesson in Newtonian physics.

The peak season for firecrackers was, naturally, the Fourth of July. That was when somewhat shady entrepreneurs would unveil the latest wonders smuggled in from our dread enemy, China. We would exchange a large fraction of our carefully horded, hard earned, lawn mowing dollars for these treasures — especially for the expensive, but prized, cherry bombs.

We acquired these munitions in July, but we expended them throughout the year — on special occasions. The last year my buddies and I went trick-or-treating we encountered the perfect, never to be topped, opportunity. Nestled that night under some trees in a quiet upper-class neighborhood was a Sheriff's Department patrol car, its occupants safeguarding the street's tranquility. Before they caught sight of us we quickly backed away into an adjacent side street to plan our tactics. It was agreed that two of us would engage the cops in conversation while the other two would stealthily make their preparations. We did not know it, then, but we had just reinvented the classic military diversion. I rather enjoyed the conversation with the polite young policemen, but when the signal came that all was ready we quickly bid our farewells, sauntered back up the road and darted behind some bushes.

*BANG!* The cherry bomb in the tailpipe blasted a yards-long flame from the car's rear end. In involuntary muscular reactions, the cops smacked their heads into the roof of the vehicle. One of my companions later swore that he could see dents in the steel top. I don't know about that. What I do remember is that we quickly evacuated the area. We needn't have worried, though. The cops were laughing too hard to pay us further heed. We got them good, and they knew it.

# THE BULLY

THE MODEL AND SEEDLING OF EVIL is familiar to all — the young bully.

When I was in high school I suffered a crisis of faith. The school was Catholic, and not being a cynic, I believed that what I was taught was God's Truth — well, mostly. In particular I believed in turn-the-other-cheek Christianity. My problem was, probably inevitably, the school bully.

He was not the first bully I had encountered. As a newcomer entering third grade in mid-term, I was naturally the target for the third grade bully — a big kid, a full head taller than I. But, in those days I was not a pacifist. I refused to back down. *He* refused to back down. We exchanged blows. He got a bloody nose. I got a black eye. As in a multitude of stories, we promptly became best friends. And, he was no longer a bully.

But I started this story from high school. That school bully, knowing I was a pacifist, found me fair game and easy pickings. Every day he would walk past my desk and punch me in the arm. It hurt in the beginning. As my arm blackened from the bruising it hurt even more. I complained to my dad, hoping he would provide a shield. He responded that I was on my own. Indeed, I found my own solution.

First I had to dispose of the pacific nonsense that I was being taught. It simply did not work. With this disposal went much of the remaining nonsensical Church doctrine.

That metaphysical debate taken care of, I was ready. The bully walked down the aisle, as usual, and cocked his fist, as usual. I didn't give him a chance. I leaped out of my seat and aimed a powerful blow straight at his jaw.

I missed.

*Wham!* My fist plunged deep into his neck, collapsing his windpipe. He flew backward over three rows of desks and lay on the floor, motionless. Within seconds he started turning blue. My stomach flip-flopped. This was very serious.

Fortunately, after a couple of minutes he recovered. For the rest of my days in high school the now tame former bully followed me around like a puppy dog, his aggression quenched. He was a nuisance, but I could put up with that.

One night, after I finished my four years of college, I was visiting my parents when the phone rang. It was the former high school bully. I had not heard from him since those days. He wanted to know how I was doing. There was some small talk and then: "How much life insurance do you have? I have a really good plan you might be interested in!"

I had eliminated a bully and left the world a life insurance salesman. Oh well, I suppose the world is better off.

# ZOT

Schuyler Hadley Basset, the Third, was a Dandy. Always immaculately turned out, he had the Upper Crust mannerisms to boot. I don't know if he came from wealth, but he certainly cultivated the image.

Whatever his real background, he did have a sense of humor — an odd sense of humor. I will not claim that I knew Schuyler well. Our paths crossed frequently, but we moved in somewhat different, though overlapping, social circles at the newly opened Irvine campus of the University of California. Schuyler was part of the campus resident community. Like myself, *my* friends were almost exclusively commuter students.

Who had the better time of it? Well, my lady friends included *the* two great campus beauties: Raven-haired Carla, deeply intelligent, psychic, scary, sexy beyond belief. A fine arts major, for her senior thesis she painted a satirical portrait of the Art Department chairman as a playing card, the Jack of Hearts. The faculty was outraged and threatened to cancel her graduation and degree. The chairman was amused. Carla got her degree. And there was Kathy. Kathy, freckled, strawberry blond, witty, brilliant math major, gifted linguist, she filled the space around her with sunshine and laughter.

Friendships are the soul of schooling. Judy Burns became my closest friend and partner in the Great Star Trek Adventure — a long story, that — a story I told in my first book.

## THE TREK CONTINUES

Then there was Bill Harter, a Physics graduate student of great talent. Bill was particularly good at annoying the faculty with rocketing super balls and other gizmos. One day Bill showed up with a lethal looking homemade boomerang. Sixteen inches long, it was a massive, knife-edged, polished aluminum, weapon of doom. We rushed outside to try out this wonder. In those days UC Irvine had only a few buildings scattered around the inner north and west quadrant of what was to become the great campus central park. The original, newly planted, central park did not yet extend up the hill as far as the Sciences building and Science Lecture Hall, so outside this outpost of civilization at UCI were the remains of the old cowboy roamed prairie.

Bill hurled the wicked wing towards the rising wilderness. The boomerang sang as it spun away into the distance. We soon lost sight of it. We waited, and we waited, but the missile did not return. Bill, his mood darkened to gloom by his failed experiment, had already turned away in disgust when I caught sight of a sun-flashed glint far away. Sure enough, the boomerang was returning. In seconds it was all too obvious that this dangerous implement was headed straight towards us. Bill and I sprinted sideways, just in time, as the very lethal weapon sliced through the air where we had been standing and buried itself deep into the rock-hard, sun-baked adobe. Of its sixteen inches, only half an inch remained shining in the sunlight. Happy as a lark, Bill disappeared into his lair and started doing aerodynamic calculations.

There were other pleasures as well. Irvine lies close by Newport Harbor and several primo surfing spots. So, much of my leisure time was spent sailing, surfing, and, of course, partying. Then, too, I had full access to a good university

## ZOT

general library, and I loved to read. Thus, not all my time was spent grinding through physics texts and equations.

It is well attested that Schuyler was having a very good time, too, though his world was quite different from mine. In one enterprise, though, we had a common goal. The goal was to promote a certain new mascot for the campus.

A mascot seems like such a small thing, such a silly, childish thing. In reality a mascot — the *right* mascot — becomes the spirit, the soul, of a university. This I remembered well from my undergraduate attachment to Oski, Cal's Golden Bear. Strip away this talisman and the institution becomes a gray place, a place mired in apathy.

Irvine, in its earliest days, was still a shell of an institution. It was yet to find its soul, its mascot. We all felt this, so the search for a guiding spirit soon became intense and not without controversy. In the first weeks and months after the university's door opened many potential mascots were proposed: *Eagle, Unicorn, Golden Bear, Roadrunner, Golden Bison, Seahawk* and, appropriately, *"none of the above."*

Golden Bear was ruled out — Berkeley already had that. Eagle, Bison, Seahawk were too prosaic for the irreverent temper of the times. Unicorn and Roadrunner had possibilities but did not seem quite right. Besides, all these proposals violated the rule. The University of California *had* a rule about this, you see. According to this rule, the mascot of a University campus must be a *Bear*. Of course, the fact that some other U.C. campuses already had non-bear mascots — Mustang and Gaucho, for example — really did not matter. At least their mascots were dignified. The mother administration in Berkeley was *not* going to tolerate any further deviation from *The Rule*.

Schuyler had his own opinion about this, and Schuyler's opinion quickly gained traction. To be sure, Schuyler's proposal most certainly violated the Berkeley Administration's rule. Schuyler wanted an *Anteater* to be the mascot. Not just any old anteater, either, this Anteater was a manifestation of Johnny Hart's popular cartoon character.

The notion of adopting The Anteater did not originate with Schuyler. Historians attribute the original idea to two good-old-boy water polo jocks, Pat Glasgow and Bob Ernst. But Schuyler was quick to pick this suggestion up from his friends and carry it to the wider university community. A substantial minority of the student body, including most of my associates, was instantly enthusiastic. The majority, however, was horrified at this mocking of academic traditions. The Berkeley administration was adamant — *no Anteater!*

Schuyler was not to be deterred. He enlisted the cooperation of Chancellor Dan. The Chancellor, Daniel G. Aldridge, was a tall, pleasant, sharp-witted man who was highly sensitive to the temper of the times. Chancellor Dan quickly became an enthusiastic supporter of the proposal. Schuyler also approached Johnny Hart. Hart was delighted and proceeded to create logos for the campaign. Then Schuyler did the unexpected — he buried himself in the library and actually did some research.

What Schuyler discovered, through his surprising diligence, successfully pacified the Berkeley Administration — well, sort of. It seems that, in South America, a species of anteater is actually a member of the bear family — the Ant Bear. Thus, The Anteater became marginally acceptable to the Establishment. This discovery certainly helped, but the game was not yet won.

## ZOT

The Official History credits the following incident as the factor that turned the tide (Who writes these histories, anyway?) — Schuyler had elected himself the University's Cheer Leader. Self-election was not difficult in those early days. With such a small student body, if you wanted the job you got it. About the only competitive sports going at the time were Sailing (which I was later involved with), Crew (which Schuyler was a member of) and Water Polo. It seems that at a water polo match, against Cal Poly Pomona, Schuyler introduced the fearsome war cry of ZOT (the sound of the strike of Johnny Hart's anteater cartoon character). This had such a terrifying effect on the visitors that UCI won the match by the surprising score of 22 to 6.

Actually, there was much more to the victorious adoption of the Anteater. Many participated actively in the propaganda blitz, including yours truly. And who knows which factors really swung the balance so that the University of California, Irvine became known around the world, with some hilarity, as the Anteaters.

Quite possibly I may have helped tip the balance. Aside from favorably talking up the idea, I wrote a little ditty, a poem, which enjoyed substantial circulation at the time. If you will allow me, I will recite it for you.

Aahh, hurummphh. (That's me clearing my throat.)

> Time was when Ants,
> Those miscreants,
> Strode *boldly* cross the plain.
>
> But there *evolved*,
> A beast involved,
> To save us from this bane.

## THE TREK CONTINUES

He *cannot hide*,
Inflated pride,
Though justly earning fame!

For Ants *walk not*,
When he goes ZOT,
Anteater is his name!

*Peter the Anteater™*

# FAILURE

## PART 1: ACADEMIC PHYSICS

WE'RE ALL SCREW-UPS. We rarely get things right the first time. This is true no matter what field we choose in life and in work. And particularly true in my profession, engineering physics.

The professional life of an engineer or a physical scientist in fact is one of repeated failure. Competent professionals in these lines of work become masters of the art of diagnose and fix. Advanced devices and concepts are *evolved* through error and correction. They are not conjured up in a state of perfection.

This said, it is always painful to fail. It doesn't matter how many times in the past you have failed it is still a pain. But I've learned that if you simply add each failure to your learned lessons you become more professionally mature. And, fortunately, you do get used to making mistakes.

Take my life. I have failed many times, in a myriad of ways. Still, many of my failures have led to success, sometimes surprisingly great success. So, I am a bona fide expert at failing, but also in recovering. I began to learn this lesson while in school.

While only gifted teachers could rouse me from indifference in the classroom, on my own I had always eaten knowledge like I was starving — reading a book a day from the time I was a child. Science and science fiction were the subjects that called to me, mostly. In fact I had first dibs on every new science fiction novel that came through the door of the local library. This was the Golden Age of Science Fiction. It didn't matter where the drama took place, the stories were always about how people dealt with the unexpected and the hazardous. Many of these novels were in reality episodes from world history transported to exotic realms. Devouring volume after volume of the highly optimistic genre opened a portal to the world.

Homework? I rarely did any until I entered the university. It became necessary in this realm, especially in graduate school. There, it was clear, homework is *essential* to mastering the increasingly difficult material. In both undergraduate and graduate school, I did homework diligently — for survival, or, more rarely, because the subject was interesting.

It took many years after I left the university for me to realize that I had never learned *how* to study. It was only then that I discovered a trick: *Make it a game.* Make it fun! Make it addictive like a crossword or Sudoku puzzle. The conversion only requires a change in attitude. This really worked for me.

There is nothing strange about learning through entertainment. It's well known that play is a powerful way to enhance learning. Watch kittens or puppies or children endlessly at play. Their play is fun, but fun is also their most effective way of learning. When I was young schools often compelled the opposite. Too many schools in those days made learning into tedious work, and *work* is not fun. Perhaps things have changed. Maybe not.

Despite my lack of knowledge about how to study, I finished high school as a National Merit Scholar. Then I entered the university that September. And nearly flunked out. What a shock! I spent the second semester of my first year on academic probation. This was at the University of California, Berkeley. I simply was not prepared for what awaited me.

For example, upon arriving at the campus I failed the entrance essay exam and had to take Subject A — bone-head English. I didn't know even the basics of how to construct a sentence. During my early schooling my parents moved around a lot. With frequent moves to new schools I'd missed learning basic grammar and punctuation. By middle school English grammar was far beyond me. I was too embarrassed to ask for help, so I faked my way through the English classes. Berkeley did me the great favor of telling me where to put a period, and what a comma is used for.

English wasn't my only nemesis that first semester. The field I chose to pursue — a major in physics — presented what turned out to be perhaps the most extreme academic challenge then offered by the university. Why did I choose physics? Well, I really wanted to be an engineer. Unfortunately the School of Engineering had an admittance exam. Oh no, not another exam! I inquired about physics. No exam for physics? No contest.

Berkeley in the 1950s was the go-to school for physics. The university had managed the atomic weapons program during the Second World War. Each year the department's glamor induced hundreds of entering students to sign on as physics majors. The attrition was pure carnage. I was fortunate, among only a dozen physics students to graduate four years later.

It was in 1963 that I got my bachelor's degree from Berkeley. As students we cynically considered ourselves to be mere commodities on the assembly line of a factory that manufactured graduates. There was almost no humanity in that institution. The Physics Department was as indifferent as the rest of the university. It was clear there was a crucial difference between the way things were *supposed* to work and the way they *actually* worked.

Consider: each physics student was assigned a permanent faculty advisor to closely guide his education. I had one. I saw him exactly twice in the eight semesters I was enrolled. The second time he didn't remember me. Once each semester, at the required study plan review, I was greeted by a different "advisory" professor who had no idea who I was. The consequence of this lack of personal guidance was that I missed something vitally important to my academic future: I did not take *linear algebra*. It wasn't required.

At Berkeley there were bits and pieces of linear algebra taught in various required math courses. The Physics Department evidently thought that those bits and pieces were all that was necessary. At the time Berkeley simply didn't see where physics was headed. Linear algebra should have been *mandatory*. It is linear algebra's abstract concept of vectors in higher dimensional spaces which is essential for understanding the physics to come. Without a formal course in linear algebra I was set up to fail in graduate school.

In my last study plan review the professor looked over my grades. He was puzzled. In some of the most difficult physics courses, such as quantum mechanics, I was the top student. In others I was mediocre. The professor couldn't make heads or tails of my record. He declared that I wasn't suited to continue in physics. Graduate school was definitely out.

I didn't listen. I was determined to go to graduate school. Try and stop me! It's a family trait. Stubbornness, I mean. Stubbornness and a desire to do things my own way.

# FAILURE

My family's history as wild west pioneers on my father's side, immigrants from Sicily on my mother's made this trait a basic part of my nature. And so, upon graduation I went my own, unorthodox, way.

My way had the plan of working a couple of years before going off to graduate school. In the spring before graduation recruiters from the technology industries showed up on campus. Engineering and physical science students flocked to be interviewed. I didn't. Since I lived in Southern California, aviation technology's epicenter, I figured I could wait a couple of months till graduation and easily find a job. I was wrong, really wrong.

In consequence, every day for three long summer months I pounded the hot pavement only to face rejection after rejection. All the entry level technical jobs had been filled in the spring recruitment drive. I met some nice people and had some good conversations during the interviews, but no dice.

I was nearly desperate when I remembered one company that was in Pasadena. The company was mysterious but I knew they had something to do with technology. I drove to their Pasadena location only to find that they were now in Monrovia. So, out to Monrovia I went, and introduced myself. They hired me on the spot. The job itself turned out to be the greatest possible bit of luck.

By doing everything wrong I had done the *one right thing*. I was hired on the spot *because* I didn't follow the crowd — the company was actively looking for mavericks. Hycon Manufacturing was the child of another maverick, Trevor Gardner. In the 1950s Trevor was a *Big Deal*. He was even on the cover of *Time Magazine*. Trevor had been Assistant Secretary of the Air Force responsible for Research and Development. As such, he was the father of the Air Force's ballistic missile program, and much else of great national importance.

Trevor created a department called *Advance Design*. This was a group of professional inventors. Advance Design was to be my home for the next two years. The department's technical leader was Bert Van Breeman. Bert was a talented inventor and my first mentor. He took me under his wing and taught me the art of invention.

In this new home I made good friends, had lots of fun, solved some important problems, saved the company a bundle of money, invented a bunch of things, received a couple of patents, wrote numerous proposals, doubled my salary, and learned how to successfully lead groups of engineers. After two years in that creative hot house I was a pro. As it turned out, I was also ruined as a future graduate student.

**WHY ONLY TWO YEARS?** It was my career plan. It was obvious that graduate education was going to be essential for my success in science or technology. I figured that two years of work would finance multiple years in graduate school. To make it all happen I had to return to the university before I got too old and too well established. Hycon blessed my plan.

I had gotten excellent scores on the Graduate Record Exam. *That* should open all doors to graduate school! An interview with the Chairman of Physics at UC San Diego dispelled that illusion. Asked why I wanted to study graduate physics I said I was interested in its applications. Wrong answer! I should have said that I loved the *beauty* of physics. Oh well, scratch UCSD.

# FAILURE

I *was* accepted at UC Riverside. Good! But then, at a party, life propelled me in an entirely unexpected direction. That evening a young lady, knowing that I had studied physics, introduced me to a new arrival. Bill Wagner. She said he's the Tolman Fellow in Physics at Caltech. I was impressed. The Tolman Fellowship is one of the highest honors a physics post-doc can receive.

Bill had just accepted a professorship at the soon to be opened Irvine campus of the University of California. He encouraged me to apply there. He even arranged for me to meet with Ken Ford, the physics chairman. The interview went fine and I was accepted as a graduate student. Even better, Professor Ford hired me to be his teaching assistant.

I was about to become part of something that was still being born.

The formal inauguration of UCI was a most memorable delight. The evening was a perfect late summer gift that only Southern California can give — soft, warm and gentle air beneath a sparkling sky. The reception was held on the terrace between the library and the commons. All wore their finest. On the balcony of the commons a brass choir, dressed for the renaissance, serenaded us with Gabrieli canzonas. Delicious aromas, the offerings of the finest chefs, drew us to the buffet. Stimulating conversations enlivened the scene. I was entranced. This new university was going to be a wonderland. I so looked forward to the beginning of instruction.

Innocent me. In reality, I was about to enter an exotic and sometimes treacherous environment. As an undergraduate I was just another anonymous widget in an academic widget factory. Pass the courses, do the labs, and you are blessed with a degree. Graduate school is completely different. Graduate school is about human relationships. It is a mine field for the unwary. Students are expected to know their lowly place. Some egos among the faculty are extremely fragile and easily offended. Being the protégé of the wrong professor can be fatal. In short, graduate school is all about politics.

My first year at UCI was very busy. In addition to physics classes I was a teaching assistant in Ken Ford's freshman lab and Julian Feldman's computer science lab. I also became involved in student politics.

UCI opened in September 1965, the year after Berkeley's Free Speech Movement. Academia was astir. Irvine, being the energetic new kid on the block, was committed to reform. Genial Chancellor Dan Aldrich gave us students his full support. I was elected to the Constitution Committee and acted as its chairman. Our job was to design the student government. In that role I also became student advisor to Chancellor Dan and to the Dean of Graduate Studies, Ralph Gerard. Dan Aldrich lobbied the Regents to add student members to the Statewide Academic Senate. I was selected for the position, representing the UCI student body. I was a busy bee.

Graduate school was far from being all grim and grindstone. UCI is situated in one of the premier garden spots of the world — adjacent to the sailing paradise of Newport Harbor and the Pacific Ocean, and near some of the world's finest surfing breaks. This was a great place to play. I was young. My friends were young. Play we did! And party, too.

Being young I still had much to learn about diplomacy. The first hint of trouble occurred at a get-acquainted reception at the home of a young physics professor, Alexei Maradudin. The two of us were having a friendly chat about the aesthetics of art when he said something that I thought was silly. I thought so and politely said so. Maradudin took it badly. We parted, never again to have an informal discussion.

But we were to see each other frequently during the coming year. Maradudin taught mathematics of physics. This is a difficult subject and I was already in trouble with him. I studied hard and surprised the professor with my growing understanding of the subject. Maradudin was an excellent teacher and that helped me a lot. Without realizing it I had turned the homework into an entertaining game. This game was to defeat my adversary, the professor. My progress in this contest was signaled by Maradudin's expression when we met in the hallways. He always had a bit of a sneer on his face when we passed each other. As time went on, and he realized that I was mastering the material in his course, the sneer gradually moderated until it had almost relaxed out of his expression. Almost.

While I made solid progress in Mathematics of Physics, Quantum Mechanics proved to be a nightmare. My friend Professor Bill Wagner (I'd had the opportunity to have many conversations with him over the years since we'd met) was the teacher. I had no idea what he was putting on the blackboard. The notation was completely alien. The ideas made no sense. The course did not, except in the most superficial way, resemble the quantum mechanics that I had mastered as an undergraduate. Obviously, I had missed something important.

Worse, the give and take that was normal in other graduate classes was absent here. The problem was Bill's physical handicap. He suffered from cerebral palsy. Problems at birth had mangled his ability to speak. I hesitated to ask him questions. Ask a question and he would stop, consider the matter at length, then deliver a beautifully crafted, but terse, response. His use of the English language was masterful, but it did slow things down a lot.

Professor Wagner was schooled in the Caltech tradition whereas I was the product of Berkeley. The two schools, in those days, were very different. Caltech offered the path to the future. Berkeley conveyed the legacy of the past. Caltech taught Heisenberg's vectors in abstract space. Berkeley presented Schrodinger's waves in physical space. They really are the same subject, but in two completely different mathematical languages. I was literate in one but not the other.

Bill had recommended I begin with Richard Feynman's famous quantum mechanics lectures. I didn't listen. I began instead with Bill's other suggestion. In order to learn the new language he also advised that I read Paul Dirac's classic text on quantum mechanics. This I did, very carefully. It was Dirac's notation that Bill was putting on the blackboard. I found Dirac's ideas difficult to comprehend. Reading his work first wasted a lot of time. In addition to my attempts to master both Feynman and Dirac I found that I badly needed a deep understanding of linear algebra.

My real problem was that I was caught in the middle of a major paradigm shift in theoretical physics. It simply was bad timing for me. Berkeley had locked itself into wave mechanics and the dead end of S-matrix theory. Caltech was on the road to Quark theory and the Standard Model. It didn't matter that I was a victim, not a villain, the pain was the same.

Somehow I passed the course, but I was making slow progress with this new version of quantum mechanics. Then, just at the point where I was beginning to get the hang of things, Bill left. He was going to USC to establish the Laser

# FAILURE

Institute. Obviously Bill had been negotiating for months. It seems evident that he was unhappy at UCI. Bill's leaving increased my difficulties. I no longer could get his help with my struggle to master quantum mechanics. Only Professor Maradudin might have helped, but that was obviously out of the question. I was on my own.

As Bill was leaving I wished him well. He did do well. In time he became Dean of Sciences at USC.

Next step for me: Ken Ford's Freshman Physics Lab. It started out routine enough. Unfortunately, after a few weeks it turned into a political disaster. The basic lab was well designed. Professor Ford had given a lot of thought to the various experiments. The equipment was already set up when I first entered the lab as the instructor. I did a trial run on each experiment. Everything worked fine until I got to the drop tower. There was a problem here. It proved to be a major problem, not just for the students, but for me as well. The difficulty was that the tower was rickety, it flexed too much.

The task, using the drop tower, was to measure the acceleration of a steel ball as it fell the length of the tower. It was a recapitulation of the monumental experiments that Galileo had done half a millennium before. To slow gravity's acceleration Galileo had rolled balls down an inclined plane so that he could use a water clock to accurately measure the time taken between locations. Our drop tower experiment measured gravity with a free-falling ball using electronic wizardry to record elapsed time.

Spaced down the length of the tower were sensors connected to binary counters. Each sensor consisted of a light and a detector. When the ball passed in front of the detector it eclipsed the light source and shut off the counter. All the counters started measuring the pulses from an oscillator the moment the ball was dropped by an electromagnet.

Neat idea. It simultaneously taught the student modern digital measurement techniques, the binary number system and the acceleration of gravity. But it didn't work in practice. The tower flexed so much that the ball seldom fell directly in front of various detectors. The sensing and timing failed. The students taking the lab were unhappy with the constant failures and were starting to get cynical.

The problem was simple, but unfixable. The original tower was constructed of pierced angle iron. These iron struts were fastened together with bolts. The *holes* the bolts went through were larger than the diameter of the bolts. So there could be lateral movement in the fasteners and the struts. At one of the staff meetings I mentioned the problem to Ken Ford. He wasn't pleased. He directed me to tighten the tower bolts and get on with it.

I knew the existing tower couldn't be fixed so I designed a new structure to be constructed from screwed together two-by-four pinewood studs. This tower would be very rigid. After sensor alignment, reliable measurements would be guaranteed. When I presented sketches of this design at the next staff meeting Ford was furious. Design was not the job of a *mere* student. I was fired on the spot. I had made a permanent enemy of the chairman of the department. He did keep my sketches, though.

The next year, as I walked by the lab, I saw that the drop tower was now my design. This gave me considerable satisfaction. Welcome to the world of engineering, Professor Ford.

In the second year I had to prepare for my PhD qualifying exam. I had already mastered Classical Mechanics and

# FAILURE

Mathematics of Physics, which I needed for the exam. So my major classroom focus was on the Electricity and Magnetism lectures. I was also spending a lot of time on my independent study of quantum mechanics, which I still had to get on top of. I abandoned student politics to concentrate on my studies and to work in Professor John Pellham's low temperature physics laboratory. In addition, a completely different science, Psychobiology, caught my attention, and some of my time.

JOHNNY PELLHAM'S LABORATORY WAS A MARVEL. Before construction of the Sciences Building had started, Pellham was busy developing his new lab. The working part was on the ground floor of the building. Below that, in the building's basement, Pellham installed a large capacity high vacuum pump and machinery to liquify helium, a necessity for most low temperature research. Piping from this machinery led up to wall outlets in the working lab. It was only necessary to plug into these outlets to get high vacuum and liquid helium.

I worked in the lab for most of a year. The lab was not just intended for Pellham's own experiments, it was also a teaching lab. My job was to duplicate, on a much smaller scale, the machinery in the basement. This meant building an ultra high vacuum chamber out of available odds and ends. My vacuum chamber created greater emptiness than interstellar outer space. No big deal, I was only learning standard laboratory practice.

We later used my chamber, together with some custom glassware, to liquify helium and make a Bose superfluid – this is a liquid you can put in motion and the motion never stops, as long as it stays cold and the apparatus stays intact.

## THE TREK CONTINUES

A superfluid does all kinds of weird things: It climbs walls. Spin up a vortex and the vortex never slows down. You can put a sound wave into the fluid and the sound will bounce around the chamber forever. Superfluid properties are truly uncanny.

Superfluid helium happens naturally when you drop the temperature very close to absolute zero. So we made some in a large intricately configured glass walled chamber. But playing games with this tricky liquid was not our intention. We were after one of my unorthodox ideas: I understood that hydrogen atoms pair up electrons and protons to make Bose particles analogous to Helium. I speculated that it might be possible to dissolve hydrogen in liquid helium to make a new kind of hybrid superfluid. Pellham thought the idea was novel and intriguing so we put our glass blower to work fabricating the apparatus.

After weeks of preparation the day arrived when we cooled the liquified helium down to near zero temperature. Time to squirt in some hydrogen gas. What we got was a beautiful miniature snowstorm. Tiny flakes of solid hydrogen drifted slowly down to form a white fluffy field at the bottom of the vessel. The experiment was a failure. We had discovered that hydrogen does not dissolve in liquid helium — it freezes out. However, the experiment was also a success in that this had never been done before and we therefore had put another small brick in the ever-growing edifice of mankind's knowledge.

While I was working with Professor Pellham he introduced me to Keith Watson. Like Pellham, Dr. Watson had been a member of the physics faculty at Caltech. He was now a successful entrepreneur and industrialist. When I met him he was Vice President for Research and Planning of the medical supplier Becton Dickinson. Keith was supervising several of BD's specialized high tech subsidiary companies. Later, Keith left BD and founded an extremely successful technology investment company. In that endeavor he had the golden touch and accumulated a pile of gold.

Keith hired me to do assorted tasks. All of them were fascinating but not time-consuming. In effect, Keith created for me a fellowship to finance my years in graduate school. I had found another real world mentor — a very entertaining one at that. My fellowship income lasted until 1969 when Congress passed a substantial increase in the US capital gains tax. Risk investment was going to become unattractive with the higher tax rate so Keith folded his investment business. He headed off to do cancer research.

One day Professor Maradudin came into Pellham's lab full of enthusiasm. He burst out with the idea that liquid hydrogen might be mixed with liquid helium to make a two part Bose condensate. Pellham responded that I had earlier proposed the same notion and we had already done the experiment. Unfortunately, Pellham explained, our result was a failure. Maradudin had the oddest expression on his face when he heard this. He abruptly turned around and left.

Toward the end of my experience in Johnny Pellham's laboratory something happened that sent shock waves through the physics department. Ken Ford told Pellham to clear out his equipment — Johnny had spent years building his laboratory. Ford wanted the space. Ford told Pellham he was going to have to create a new laboratory in the yet unbuilt Physical Sciences building. Professor Pellham's research was aborted and so were the years of work building the lab. You can imagine how Johnny Pellham took this diktat.

The department's faculty was appalled. Not only was Professor Pellham highly esteemed in the world of physics, this affable man was very hard not to like. Unfortunately sufficient members of the faculty were insecure and craven enough that Ford got his way and Pellham was evicted from his lab. If the department had not had factions before it certainly did now.

At the end of my second year it was time for me to take the PhD qualifying exam. The exam was divided into four parts: Classical Mechanics, Mathematics of Physics, Electricity and Magnetism, and Quantum Mechanics. These bedrock courses were the only compulsory ones in the department. The rest were elective specialty courses. The faculty was rational enough to recognize that students were not going to be equally strong in all four subjects. As a result they softened the requirements for qualification. As I remember, a student was qualified if he passed three of the four sections and exceeded a lower threshold on the fourth.

How did I do on the qualifying exam? I passed. And then I failed. How could this happen? Johnny Pellham gave me the low down. His description was later confirmed by other friendly members of the faculty. Remember that one had to fully pass three sections and be above a threshold on the fourth? Well, I easily passed three sections and, being above the threshold, came within a few points of fully passing the section on quantum mechanics. I was not happy with my performance in quantum mechanics and feared that I had failed, but hoped otherwise. Unknown to me at the time, by the three and threshold criterion, I had actually *passed* the exam. Ken Ford, among others, was not pleased at this. He sequestered the physics faculty in a closed conclave to discuss my case. It was to be a long afternoon, and an even longer night for the faculty.

The conclave's debate over my situation was bitter. I suspect that along the way I had unintentionally stepped on a few extra toes — in addition to Ken Ford's. According to the established rules I had passed the exam. That was completely unacceptable to some of the faculty. Eventually the balance swung against me and I must be failed. But still, I had passed. Finally a solution was found. Since the rules for passage had not been published the rules could be, and were, amended to eliminate the fourth section threshold. I was duly failed and my failure was entered into the record. Unfortunately, that raised a dilemma. Now, *all* the students taking the exam had failed. This mass failure would attract very unfavorable attention from the wider academic community. What made the situation worse, the exam results had to be posted first thing the next morning.

What to do? Somehow the rules had to be changed back so that other students could pass. Unfortunately for the tired professors, the countervailing rule change could not take place on the same day. To make the second rule change quickly enough to meet the next morning's posting, the faculty had to remain sequestered until after midnight and the advent of a new day. And so it was done. This amazing ex-post-facto juggling of the rules had saved the public reputation of the Physics Department — but not its honor.

Johnny Pellham was still steaming when he told me this tale. What Pellham said next was completely unexpected. It was Alexei Maradudin who had kept the debate going. It was Alexei Maradudin who had argued the most strenuously for my being passed. Maradudin was my *champion*, not my nemesis.

I wasn't just surprised, I was stunned — and *ashamed*. It was *I* who had been wrong about Professor Maradudin. And badly wrong, at that. I had completely misjudged the professor. This also meant that I had missed many opportunities to learn from a good man. But now the damage was done. I was shown the physics department door, my consolation prize being a Master's Degree. It was 1967.

Subsequently I learned that the next year Alexei Maradudin had replaced Ken Ford as chairman of the physics department. It was the beginning of his long and very distinguished career as an administrator as well as a scientist. Maradudin began a reformation in the physics department and things soon settled down there. Ken Ford left the university two years later in 1970.

# FAILURE

## PART 2: ACADEMIC ENGINEERING

AFTER WASHING OUT IN THE PHYSICS DEPARTMENT I wasn't finished with graduate studies. I had another idea — Engineering.

I didn't just walk across the campus and join the School of Engineering. They asked me to take an exam to be admitted. That's the School of Engineering, of course. I ran into the same thing at Berkeley. Oh well, take the exam and be done with it. The exam was, in many respects, similar to the one I had just taken and failed. It had the same three bedrock subjects of mathematics, classical mechanics and electricity and magnetism. I was relieved it didn't have quantum mechanics. It had a grab bag of problems in different subjects instead. Many of the problems were more difficult than one would expect for an entrance exam. This included one killer problem in mechanics that definitely was intended to separate the men from the boys.

And boy, did it! Of course they didn't expect me to solve it. The question was there just to see how I would approach the problem. For you tech weenies out there sink your teeth into this one: Picture a board suspended some distance below the ceiling. In the board is a hole. Attached to the ceiling is a spring. The spring can swivel in a circle around its ceiling attachment. A chord is attached to the far end of the spring and is passed down through the hole. Below the board the chord is fastened to the surface of a steel ball which can spin in all three directions around the off-center attachment point, it can also swing freely, and it can bounce up and down with the flexing of the spring. The task: write the equations of motion for the ball and solve them!

Yeah, right. I saved that one for last.

To my relief, I passed the exam, and without difficulty! The killer question was easy for me because of Dave Pandres' superb course in Classical Mechanics back in the Physics Department. All I had to do was figure out the Lagrangian, insert it into the Action Integral and apply Euler-Lagrange's equations of motion. To my amazement the result partitioned into coupled harmonic oscillators which could be solved individually. Piece of Cake.

I was duly accepted into the School of Engineering. Upon my acceptance a surprise was in store. The exam I had taken was, in fact, my PhD qualifying exam in engineering. I was now accepted as a Doctoral Candidate as well as an admitted graduate student. Pleasant surprise, that!

Since I was a physicist rather than an engineer it was necessary for me to take a substantial number of engineering courses. Two years of courses as it turned out. Because of the convergence between modern engineering and physics these courses were all heavy in advanced mathematics and felt very similar to physics courses. Some were pure mathematics, delving into interesting mathematical realms that were completely new to me.

Perhaps the most interesting, and later useful, course was Control Systems. This course included a laboratory. Once again I could get my hands dirty working with real physical objects and seeing real physical things happening according to prediction. The lab was fun. The course was enlightening and entertaining. I was enjoying being back in the classroom. Even better, I was able to apply what this course taught me to my dissertation research.

I also supervised the electronics lab as a teaching assistant. As expected, the students always had trouble getting their transistor amplifiers to work. I would walk by, examine a breadboard layout, jiggle a wire or two and all of sudden the thing would work. "How did you do that?" the student would ask in wonder. "It's magic," I would reply. It really was magic because I didn't have a clue what I was doing. It was something that my boss, Bert Van Breeman, had shown me back when I was working at Hycon Manufacturing, right after getting my bachelor's degree.

As a hobby I was experimenting with transistor amplifiers, trying to make them work. I followed the diagrams and did the necessary calculations, but to no avail. What happened, mostly, was destruction of the transistors. And transistors in those days were expensive. I mentioned my problem to Bert. The next day he handed me a small jar of very expensive transistors from his personal stock, and told me to use them. I protested at the cost to him. He told me it was okay, they were surplus. He said I would burn up a bunch of these transistors but when I had reached halfway through the bottle things would start to work and I would never again fry a transistor. "It's mysterious," he said, "but it always happens that way." He was right. I blew up half the transistors but when I reached the half bottle point I could no longer make something that didn't work.

# THE TREK CONTINUES

Most graduate students spend considerable time trying to decide where to specialize. I was no different. The path that took me to my doctoral research was the consequence in part of campus geography.

In the beginning UCI was a small school of fewer than sixteen hundred students. It was smaller than many Southern California high schools. In time the student body would grow to today's more than thirty-seven thousand. In those early days UCI did not require extensive facilities. As a result all the sciences were housed in one building. For two years as a physics graduate student this building was my home. For me this physical concentration was significant because the Physics Department offices were located on the same floor as Psychobiology. Every day I would walk by a number of the psychobiology labs. It was inevitable that I would become acquainted with some of their graduate students and the psychobiology faculty.

Psychobiology at Irvine focused on experiment rather than theory and those experiments were sometimes more than a little hard to take. One of my friends in the department, a post doc, took me into the lab where he was working. I was only slightly bothered looking at an archaic mudpuppy with wires trailing from its head to an electronic instrument. The wires were connected to electrodes monitoring individual neurons inside its brain. I observed with interest how these primitive animals reacted to various stimuli, but I didn't much like this kind of experimentation. Nevertheless, what I learned helped save my life when I later had a too intimate a romance with a Nile crocodile.

What I really hated was what these people were doing to a room full of caged cats. Their brains were similarly wired up for experiments, with the brain probes attached to a socket glued to the skull. Soon enough they would be plugged in to the instrumentation. It broke my heart to hear the plaintive cries of the caged kitties. What was being learned was valuable, but at what cost?

These encounters with the psychobiology professors and students opened up a fascinating new world for me. A century of intense scientific investigation had compiled a large library of discovery. I acquired a copy of the standard undergraduate textbook and plunged in. I was so involved in this material that I soon worked my way up to the professional literature. The department's colloquia introduced me to the latest research and to the field's leading figures.

I started thinking about one of the discipline's greatest unsolved problems: What exactly is memory and how does it work in the brain? Somehow memory must involve the reconstruction, or replaying, of previous neural activity. For example, when you see a written word on the page this creates a specific pattern of neural activity in response to just that *picture* of the word. This visual neural activity pattern somehow also creates the *sound* of the word in your mind — the sound being a different pattern of neural activity. It's as if the *picture* of the word is the key which unlocks the sound activity. You recorded this sight-sound association when you were learning to read.

How might this reconstruction of neural activity work? This is where my knowledge of physical optics came into play. One day it suddenly occurred to me that what is taking place in the brain might be analogous to optical holography. Holography is the technology of reconstructing complex waves of light. That flash of insight changed my life. I started to call my idea *neural holography*. Others were independently thinking along similar lines and were also calling the subject neural holography, so that name stuck.

Optical holography is the art of capturing *all* the essential information in a wavefront of light. With this information, at a later time a new wave of light can be created which is essentially identical to the original.

When you look at an object you really don't "see" the object. What you see is the wave of light that was reflected from the object. Before that wave enters your eye it is all crinkled up with information about the object. This crinkled wave is a *source* wave.

To make a hologram you place a piece of sensitive photographic film in the path of that wrinkled source wave. A *reference* wave that impinges on the source wave makes the wrinkle pattern visible. This wrinkle pattern is then recorded on the photographic film. Later you can pass the reference wave of light through the developed film. Out comes a wave that is identical to the original source light that was reflected from the object. We say that the reference wave of light *reconstructs* the source wave.

When that later wave enters your eye, you will "see" the recorded object as if it were still physically there. It is in three dimensions and appears *precisely* the same as the original. But you really see only the reconstructed wave of light, not the original object. Everyone who encounters a true optical hologram regards it as a kind of magic. It *is* scientific and technological magic! Dennis Gabor well deserved the Nobel Prize for his invention of holography back in the 1940's.

I postulated that a memory is recorded when a specific *reference* "wave" of neural activity interacts with the neural response pattern resulting from a simultaneous *experience.* An association is thereby formed between the reference neural activity pattern and the experience pattern. Later on, when this specific *reference* pattern of neural activity is again present, it reconstructs the experience pattern of neural activity.

One really important idea that comes out of neural holography is that a specific memory is distributed over millions of neurons just as the light wave's phase map is spread over the photographic film. It is *not* localized to one, or a few, neurons as once had been believed. A distributed memory trace had been postulated two decades before by Donald Hebb, but no really plausible mechanism had been offered as to how it might work. Holography offered a promising explanation. It was definitely worth further investigation. So I decided to write my doctoral dissertation on this topic.

The School of Engineering accepted neural holography for my research. They couldn't help me much with it, though. The whole concept was too new. I was pretty much on my own. My third year in engineering I spent writing my dissertation. At the end of the third year I had to face my Oral Exam for acceptance of my dissertation and graduation.

The examination panel was composed of representatives from the local faculty and some strangers. I started my presentation with an exposition of the scientific background. This set the scene for the introduction of the problem of memory and my proposed solution. Then I got into the details of my ideas and where they were novel. After that came questions. Most concerned the material I had presented. So far I was doing fine except for one detail that I found very troubling. I recognized that a key equation was wrong. In this equation I was sometimes dividing by zero – such a fundamental mistake! This sent a wave of confusion through my mind. Why had I not seen this before? Probably because the problem was so fundamental the solution was ugly, and I didn't want that. In fact nobody on the committee caught it either, and it took me years to find the right solution.

Then came a totally unexpected question from one of the examiners. The question was out of the blue. It was intentionally so! It was meant to trip me up to see how I would perform under pressure. "Define linearity," I was asked. I responded with a garbled answer. The examiner asked the question again. Somewhere in the past I must have run into the answer he wanted but I just couldn't pull it out of my memory. I got confused and a bit panicky. We went on. Completely different questions were now being asked, but by this time I was stumbling on through a mental fog. It was as if the question about linearity, and my concern about the rogue equation, had flooded my mind with powerful jamming signals. Eventually the ordeal was over and I fled.

During the later debriefing I was told what I had to do to finish up. The requirements really weren't very demanding. I certainly could do the tasks in a few months and get my degree the next year. I was being treated very gently. I didn't listen. I felt so embarrassed at my performance that I just wanted to hide.

It wasn't just my embarrassment at the unexpected question that had paralyzed me. The real problem was that wrong equation. I realized that there was no way I would publish a dissertation until I had found a solution to the problem that I had just discovered.

I decided to take a leave of absence on campus. The Graduate Dean, Ralph Gerard, was still very interested in

## FAILURE

my ideas and provided me with a grant to use the university's science computer. That had to be done at night because of the daytime demands on the machine. But working all night on the computer, and during the day on the theory, was exhausting and I wasn't making progress. After a few months I could do no more. Besides, since Keith Watson had folded his investment business my bank account was rapidly diminishing. I threw in the towel. My second attempt at a PhD was over. There would be a third.

# FAILURE, AND SUCCESS

## PART 3: THE REAL WORLD

Leaving graduate school when I did was the second-best decision I ever made. Marrying my dear Sarah was the first. More about that in the stories about Sarah.

When a student seeks a PhD he hopes to become the apprentice to a master. But master practitioners in the world are rare. More than likely the student will be supervised by a journeyman. Few will understudy with a master. Only a very select few will become the apprentice of a grand master. At Irvine I was privileged to work for Johnny Pellham, a grand master of experimental physics. By leaving the university at precisely the time I did it was my very good fortune to learn, in the coming years, from no less than four additional grand masters. For me the stars were aligned.

Graduate school behind me I headed up to the top of Jamboree Hill in Newport Beach. I was about to slip through a very narrow window of golden opportunity. There, a few summers before, I had worked at the Aeronutronic division of Ford Motor Company. My old boss had an idea about what to do with me. He wanted to introduce me to a fellow named Bill O'Neil. My old boss said that I was lucky to have come when I did because Bill just happened to be staffing up.

As we were walking down the hall one of my colleagues from the previous summer job came out of a side corridor. He was glad to see me and wondered why I was there. I mentioned that I was being taken for an interview with Bill O'Neil. My old comrade suddenly became very agitated. "Watch out for Bill," he said. "He's really *scary*, and dangerous too. The man's a genius!" As it turned out, Bill really was a genius.

I thanked my friend for the caution. My old boss and I continued on to Bill's office. Five minutes after the introduction I had nailed down the job. Then Bill and I spent the rest of the afternoon laughing about everything under the sun. That day a friendship started that lasted nearly half a century until Bill passed away. Bill was the first of my new mentors.

## The Bill O'Neil Experience: You Don't Need A University To Learn From A Master

WORKING FOR BILL O'NEIL WAS LIKE returning to the heaven of Hycon, the place I mentioned earlier where for two years after I received my bachelor's degree I worked in an exceptionally fertile environment. This was immediately before I enrolled in graduate school. Hycon had tapped into my creative side and trained me in the art of invention. I'd never had more fun. Lucky me!

Now, working for Bill at Aeronutronic, the constraints on my creativity again were lifted, and I had a highly productive and entertaining decade before we both moved on to other opportunities.

Sometime during my years at Aeronutronic Bill sent me off to introduce myself to another grand master, Bob Pons, a one of a kind Texas raconteur. Uncle Bob, as I took to calling him, was expecting me and took me under his wing. (Uncle Bob also appears in a "System Architecture," a story in my book *From The Potato to Star Trek and Beyond*. The discipline is a powerful project management skill, and I am eternally grateful to Bill's for imparting his wisdom about the field.)

At an early age Bob had rapidly worked his way up to become Chief Engineer of the Cadillac Engine Division of General Motors. In that job Bob was responsible for the development of all of GM's engines. A few years later, he went to work for Henry Ford. Henry gave Bob the assignment of building the world's greatest racing engine. Bob moved his team to Aeronutronic for the project. As shown in the movie *Ford versus Ferrari*, Carroll Shelby's Ford Le Mans racing car used Bob's engine — the engine that could be pushed far beyond the bounds of expected destruction and still deliver.

At Aeronutronic my assignments were all over the technology map. I found that if you keep doing different things your skills will keep developing.

After almost a decade I had matured enough in my technical skills that I was ready to take another look at the subject of my doctoral dissertation, neural holography.

Upon reexamining the topic with my newly developed skills, I made some new discoveries. Moreover, I was able to solve a theoretical problem that had frustrated me during my oral exam. I could now derive the critical equation that I needed. With these new ideas in mind I had a chat with the local representative of the Office of Naval Research. He liked what he heard and gave me a contract.

The work was a great success. My most important new idea was to put a digital representation of a hologram inside a negative feedback loop. When I did this out popped solutions to many problems in psychobiology. I could now explain many perceptual phenomena that were well known but not understood. My Navy contract officer was delighted. He invited me to join the ONR Brain and Behavior Study Group. It was a treat to be a member of this prestigious group. I was able to get to know some of the leading figures working in the biological sciences, and remained an active member for several years until it folded.

With my neural holography discoveries in hand, I decided to finish my PhD. My dissertation would be a block buster. UCI Admissions said I had to start from scratch. I got no credit for five years of graduate school. I would have to take a full course load for years then pass a new qualifying exam. The real killer was that my dissertation must be on a *completely different* topic. It was an open insult. In other words: "Get lost!" I could have had a chat with Chancellor Aldrich. He was my friend, after all. Instead I decided to walk away. My opinion of UCI had sunk to rock bottom.

With projects like I had at Aeronutronic, and for the rest of my career, I found myself immersed in a science fiction world. It was as if I had become a character in a futuristic novel. Only, this was for real. This *was* the future. One of the major projects I worked on at Aeronutronic was the control of High Energy Laser weapons — *death rays*. I found a way to improve their pointing and tracking accuracy by substantially more than a factor of a hundred. That got a lot of attention.

The HEL project not only was fun, it aimed me directly towards meeting my next two grand master mentors. One was Max Hunter. Max masterminded the Thor rocket which evolved into the work horse Delta series of space launch vehicles. Delta rockets have put hundreds of our largest reconnaissance satellites into orbit. Another of Max's achievements was the DC-X rocket test vehicle. This testbed proved it is possible to automatically land a large rocket vertically. It was the father of today's vertical landing rockets such as the SpaceX Falcon and Star Ship. Max coached me about launch vehicle architectures.

My fourth post-academic mentor was Tom Coultas. Tom was a key figure in the development of the giant first stage F1 Saturn rocket engine and the upper stage J2 engine — engines which took us to the moon — and the Lunar Ascent engine which brought the astronauts back. He was also the systems master for the Space Shuttle Main Engine. Tom taught me jet and rocket engine thermodynamics and coached me through the early analysis of a new type of turbine engine I had invented, for which I now have a fundamental patent.

It's from this saga I learned that not only is schooling not all acquired in school, you don't need a PhD to succeed. A PhD is a very useful tool to open doors. But there are other ways to open doors — with *reputation* based on *accomplishment* being the most effective. It is not the open door that is important, it is what you do *after* you have walked through it that matters.

At this far end of my career it is amusing contemplate the might-have-beens. If I had continued in academic physics or engineering I might have had a career something like my old Physics Department nemesis and champion, Professor Alexei Maradudin. In addition to being a distinguished teacher and administrator he produced a multitude of papers about interesting discoveries in a specialized field of research.

But my experience demonstrated that I do not have the temperament for a specialized career such as this. Scientific research would not have left much room for my main talent — invention. Invention needs elbow room. I have been fortunate to contribute ideas and inventions ranging widely over the technology spectrum from optics, radar and signal processing to jet engines, rockets and satellites — with useful mathematical discoveries in between.

If I hadn't been kicked out of the Physics Department I would not have gained the knowledge I needed from the School of Engineering. If I had not failed in my engineering dissertation project and bailed out of the university when I did I would not have become the protégé of four great engineering grand masters: Bill O'Neil in aerospace engineering, Bob Pons in system architecture, Max Hunter, the great rocket pioneer, and Tom Coultas, the wizard of rocket engines.

I have had a wonderfully varied and satisfying career. All because I have, on many occasions, failed. And, once in a while, failed big time.

So you see, failure *is* an option!

## BUG AT THE BEACH

WITH LUCK LIFE GIVES YOU A FIRST LOVE. Sometimes life is generous and gives you multiple first loves. With me it was Linette, my pal in first grade. Then beautiful Becky, in eighth grade, who I could only admire from afar — she entered a convent. College presented me with spectacular Carla and radiant Kathy and dear Dianna. Then came my true life love, Sarah, whom I wooed, won, married, lived with, lost and buried.

Well, there can be other types of love — maybe affection is a better word — in one's life. How about one's first automobile? Actually not. It was my second that I developed a well-deserved affection for.

Car culture. I grew up in the heart of car culture — Southern California in the 1950's. Naturally you adopt the culture around you, so I did. During my teenage years I lived in the upscale neighborhood of La Canada and nearby Flintridge. My family was not yet prosperous so I was the "poor" kid at school. When my friends drove their shiny new cars to school I joined in the collective admiration of their beauty. But I could only admire them as abstractions. All I had was my bicycle and that would have to do for now.

Still, a bicycle was transportation and it could go places a car couldn't go. I exploited that different kind of freedom for I had the trails. Immediately behind our newly built tract home was a bridal path. This led up the hill past two equally new houses into wilderness then snaked around Gould Mesa, descending near the parking lots of the still new Jet Propulsion Laboratory and terminating in Oak Grove Park. It was this route I took when I headed on bicycle excursions up Arroyo Seco Canyon, and its trout laden stream, riding deep into the San Gabriel Mountains.

One day I was cascading my way down the JPL parking lots when I came upon a marvel. From a distance it looked like a toy, something you might find under the Christmas Tree. It wasn't, but you had to get close to be sure. It was a small, bulbous, beetle-like thing, quite unlike the streamlined, tail finned, vehicles that America was then producing. Into its back was inserted a large windup key. When you got close you could read instructional placards – one in German, the other in English. These instructed the operator to firmly hold the wheels while winding up the motor lest the wheels spin out of control. This was my introduction to the soon to be notorious Volkswagen.

The car was perfectly functional and became something of a celebrity around town. It also wittily opened the door to California's car culture and therefore to VW's acceptance into America. A dozen years in the future I was to become well acquainted with this critter.

In still developing La Canada, new cut roads were being paved and more houses built, spreading civilization ever upwards towards the looming mountains.

The first summer we were in our new house we gathered with our neighbors to construct a tall, steel reinforced, concrete block retaining wall. This, to keep three of the houses across the street from washing away in the expected heavy winter rains. The spirit of this land was still a bit of the Old West. Neighbors still helped each other out as they had done during nineteenth century barn raisings.

I helped dad with construction around our new home. We built brick walls to isolate our yard from the bridal trail and provide privacy between our adjacent houses. Dad cut a doorway into the lower wall of the house and together we laid a concrete floored basement and workshop. A carport was built after my colicky sister was born and the garage was converted into a suite for me so that I could get some sleep during my high school years.

The old house, and all we put into it, is gone now, replaced by a Persian Palace built by an Iranian import. The value was in the land not in our wood and work. That's okay. A house is a kind of tent. You pitch it to keep shelter and memories. When you leave you may keep your memories but the tent is folded. The land now belongs to someone else.

The year I finished high school my parents bought a new house in the hilly Chevy Chase district of Glendale. That was a good place for my sister to grow up – good schools, good neighbors, quiet – except for the Great Glendale Fire of 1964 which burned off our backyard and dozens of homes across the city. The fire also temporarily inconvenienced the family of coyotes that lived in the chaparral beyond our patio and shared our property with Samantha, our orange tabby cat (her job was to keep the field mice out of the garage and understructure).

In the beginning I wasn't there much, spending my time at Berkeley picking up a degree in physics. Summers during those undergraduate years I borrowed Mom's car to get to my summer jobs: electronic assembly, laying up fiberglass plastic, and one summer working as a technician at JPL.

Physics is limited as a profession without an advanced degree. But graduate school takes money and the family budget was being devoted to the education of my sister. The solution was for me to work for a while as an engineer and save up money for my further education. I stayed with my parents, and paid them room and board, during two years before I went off to graduate school.

I found a job in the Advance Design Department of an engineering company in distant Monrovia. It was a long commute in the days before freeways but the job could not have been better. This was a group of professional inventors and I was being trained as an inventor. During the job interview my unorthodox approach to the world had attracted their attention and they scooped me up. What luck!

I needed a car. A neighbor sold me his old Nash for a couple hundred bucks. It wasn't much, and it required a lot of maintenance, but it was cheap and got me to work. After a year I had enough money to buy a proper new car so I started looking around.

I was now working on military projects. And, I was learning from my new colleagues what was good engineering and what was bad (there is a *lot* of bad engineering!). Military engineering is special and expensive because *reliability* is paramount. There is nothing worse than a weapon that does not work.

I was familiar with war stories, of course. This was a period when the Second World War still heavily influenced the culture. Along the way I had read Ernie Pyle's accounts of our activities in Africa and Europe. Pyle had a particularly quirky way of reporting what he observed as a war correspondent. One of his more humorous reports described the reputations of competing vehicles: Kubel Wagen vs. Jeep.

It seems that the Germans just loved to get their hands on an American Jeep. Oddly, the GI's much preferred the Kubel Wagen that the Germans so despised. I guess the grass is always greener.

Well the Kubel Wagen was just an open topped, militarized Volkswagen. Its GI reputation for go-anywhere reliability transferred to the newly arriving post war VWs. My colleagues at work also recognized good engineering when they saw it and recommended it for my consideration. What I needed for graduate school was good, cheap, highly reliable, transportation. The VW was perfect for my needs. I bought a straight-from-the-factory one for fourteen hundred dollars.

No machine is perfect. A VW of that generation was no exception. My new car had its virtues – some very great as we shall see. Its vices were equally great and sometimes totally unexpected.

The car's two great mechanical vices were its aerodynamics and its balance. Both could kill you — at least until you became familiar with them and developed strategies and tactics to compensate.

Machines are tools with a mission. They are just as susceptible to evolution by natural selection as are animals. Airplanes have wings in front and tails at the back. So do birds. Nature discovered the reason hundreds of millions of years ago: It works best!

Automobile design is similarly driven by natural selection. A car needs to be steered. This means the front wheels must turn to control steering. An engine needs to be controlled directly, with instant response, so it is best to put the engine close to the driver — up front with direct controls. However, it is very difficult and expensive to make a front wheel drive whereby the front wheels are also doing the steering. Therefore, drive the rear wheels instead. So you install a drive shaft and a gear box down the center of the car. Voila! you have the standard car layout for the twentieth century.

But you don't have a Volkswagen. Volkswagen had a different mission than the luxury vehicles which dominated European automobile design. Its job was to provide the very most economical transportation for the masses.

Henry Ford had already done this for Americans with a conventional layout and an extremely simple standardized design. But his cars were still too expensive for 1930's Germany.

In nature there are different types of animals — different solutions to survival. And, like God in the Garden of Eden, Ferdinand Porsche created a different type of beast. Porsche discovered how to have the driver's controls for the engine and transmission be instantly responsive with the engine in the back, where the power was going directly to the rear wheels. It took some clever gear box engineering to make this happen.

But why put the engine in back? The answer is you get the largest possible passenger space in the smallest, cheapest, vehicle. Unfortunately, you also get the car's vices, as well.

Aerodynamics was a new science in the 1930's. Airplanes were becoming streamlined and more efficient. Naturally this attracted attention and *Streamline Moderne* became the fashion for architecture, furniture and automobiles. All well and good as an artistic movement but it produced some shockingly bad cars. Mom's post war Plymouth coup tended to be blown all over the highway by heavy winds. The similarly shaped Volkswagen had the same dreadful characteristic.

It wasn't until the 1980's that engineers were finally permitted by their marketing dominated management to get it right and we now have almost interchangeable *jelly bean* car designs.

I discovered this vice of the VW the hard way. One night I was driving north on the Harbor Freeway. A stiff irregular wind was blowing and my beetle was dancing around in the slow lane as I struggled to keep it under control. Alongside the freeway was a group of industrial buildings between two of which was a narrow gap. The configuration was such as to generate a wind funnel. This sudden wind blast blew me all the way across three lanes of freeway and almost into the center divider. Fortunately, it was late at night and traffic was light so the few other cars were able to dodge me. No harm done, but lessons were learned.

The VW may have been shaped like a beetle for practical reasons but, like a real beetle, it had far too much aerodynamic lift to be safe in a high wind. It tended to fly up off the road and become almost airborne. Lots of cars had that problem until it was realized that good car design requires spoilers in their shapes to push the car down towards the road at high speed. That is why modern cars look so similar regardless of brand.

Somewhere along the line the nickname "beetle" changed to "bug" and the "bug-catcher" exposed me to another VW vice — its balance. The bug-catcher is one of the exchange ramps of the Los Angeles freeway system. Normally I didn't use that interchange so I did not, at first, know about its notorious tendency to throw rear engine cars spinning off into space.

Fortunately, someone tipped me off so that I would be wary should I ever have to traverse this trap. Sure enough, on my first visit to the bug-catcher I was nearly caught even though I was being particularly careful. Why this ramp was so peculiar is unknown to this day. As I have mentioned, there is a lot of *bad* engineering in this world – particularly in road design. People get killed.

Still, one of my new car's vices was not the fault of engineering. One characteristic had to do with emotions. Years later I had a hard time convincing Sarah to consider Honda products. Her father had been a fighter pilot and was lost in the Philippines. Being an orphan, she wanted nothing to do with Japan. Eventually she came around and loved our new cars.

I understood. In the early days the VW beetle was deeply resented by many people — veterans perhaps, holocaust survivors perhaps (I later was friends with one so I gained some understanding). In any case I discovered early on just how much hate my new beetle engendered in some people.

Not many days after I acquired my new VW I was driving to my job in Monrovia. Along a stretch of road in Pasadena a large truck pulled up beside me and started pacing me. I slowed down, he slowed down. I sped up, he sped up. I looked up at the truck driver and caught the expression of scornful hate on his face as he looked down at me and deliberately turned his steering wheel in my direction. Up onto the sidewalk I went. Catastrophe was avoided but I was shaken. In the million miles I have driven around this town I have seen this kind of thing a number of times. Road rage is real.

Life is a pin-ball machine and we are pin balls being bounced around through chance encounters. I had already

## THE TREK CONTINUES

been accepted as a graduate student and was full of plans when one evening at a party I met someone who completely changed my direction. My new friend was joining the faculty at the soon to be opened Irvine campus of the University of California. The opportunity to maybe help shape something new, something you know is destined to be great, was irresistible. My friend opened the door for me and I enrolled in the first graduate physics class at the new university.

U.C. Irvine's Inauguration was held one sparkling September evening on the plaza between the Commons and the Library. It wasn't a large crowd because the university in those days was much smaller than a typical Southern California public high school.

The evening was a delight. The air was soft and warm. A brass choir, clad in Renaissance costumes, serenaded from the balcony of the Commons. The finest buffet provided refreshments. Conversation among the immaculately dressed was erudite. It was a perfect opening for the grandeur that was to follow as the then small campus was to rapidly expand to its current enormous size.

Then came the rains!

The initial campus construction had not quite been finished. The delay was caused by a brief visit from the President of the United States. The State *Powers-That-Be* just had to show off by inviting President Johnson to the Dedication.

The problem, they discovered, was that the president's helicopter couldn't land on just any ordinary parking lot. *No Sir!* No, the landing spot had to meet Air Force runway specifications. There was a mad scramble to tear up the newly completed parking lot next to the Commons and replace it with sixteen feet (someone said) of steel reinforced concrete.

Naturally supplies and schedule ran out and the other parking lots and sidewalks planned for the campus remained unpaved.

Then came the rains!

Then came the greatest virtue of my bug. The GIs weren't kidding when they claimed a Kubel Wagen could go anywhere and do anything. So could my new VW.

The mysteries of El Nino and La Nina remain unsolved to this day. Southern California is officially a desert. There are years when we get fewer raindrops than the Sahara. But then it rains. That September the rains came unexpectedly early. It rained, it rained, and it rained some more. We had weeks of drenching rain.

Rain means mud. Well, it meant mud at Irvine. No sidewalks, no paved parking near the classrooms. It meant mud. Thanks Lyndon Johnson!

You adapt. Arrive at school, or leave the dorms, take off your shoes and carry them. A backpack helps. Galoop through the mud to get to your class. Clean off your feet as best you can and put your shoes back on. After a day or two why bother. Leave your shoes behind. Everyone did. We were all in this together. It was commonplace seeing a suit wearing professor, pants rolled up, walking barefoot through the mud. Why Not? After all, the rains were early and therefore warm. There was no great discomfort walking barefoot. The events of those weeks set an informal tone for the university which still exists to this day.

With our young ladies the barefoot fashion lasted years after the rains ceased and the sidewalks were paved. If you are old enough you may remember pretty damsels traipsing barefoot through spring blossoms. That became the fashion among the young in the late sixties. Fashions start somewhere. Barefoot babes began that September at Irvine.

One morning I drove to school before the rains let loose and parked in the nearby designated, but still unpaved, parking lot. It had been a long day and was already turning dark when I went back to my car. The rain had started early and by mid-day had already alerted all as to what was about to come.

In the evening I gingerly made my way across the now muddy field, got in my car, and drove back to my apartment. In the morning, when I returned, several of the cars I had left behind were still there. Some of the faculty had been stuck there over-night and they were not happy.

The university made emergency arrangements and bulldozers were hastily brought in to haul vehicles out of the mud. I never did have a problem. No matter how deep the mud, no matter how hard those dozers had to work to pull cars back onto the paved roads, I simply got in my VW and drove off. On one particularly nasty night, just for the fun of it, I showed off by driving a complete circle around a tractor and its tow. Bad boy!

Still, some people don't learn and cars continued to be parked in the unpaved lots. The bulldozer crews made overtime for a while.

So, as you can see, there are good reasons why the original beetle was rapidly becoming legendary.

The VW was versatile. It could be adapted to fill many roles. There even was a legend that VWs could float. Who could believe that?

One person tested this by converting a Beetle into an amphibious car. That wasn't hard. Just finish sealing an almost impervious bottom and gasket the doors. With that, the car had sufficient floatation. The engine was mounted high enough to keep running. The remaining modifications were simple, too. The only significant hole in the floor were air ducts for the heater (no air conditioning in that primitive vehicle.) The publicity film showed the VW driving off the road and paddling its way across a lake, its spinning rear wheels providing propulsion.

There was a time when I was young and foolish. Really! You don't believe me? What happened one day long ago will enlighten you.

It was an El Nino year. It had been raining hard for a week. I was home from school staying with my family when I decided to drive out past Monrovia to visit one of the friends I had made at work.

From that part of Glendale you drive down into the Rose Bowl area and then across Pasadena, Arcadia, and beyond. This had been my normal route going to work. The Rose Bowl canyon had a reputation for flooding, but I had traveled the road many times and had never had a problem, even in very heavy rains. This day I did.

The rain had ended several hours before. I drove down towards the Rose Bowl and was proceeding along the familiar road when I came to a barrier that the city had set up. Beyond, in the far distance, the road was covered with a slick of water but otherwise all seemed normal. I drove around the barrier and proceeded on my way, but slowly as I approached the slick as the water was starting to sluice up around my wheels.

Suddenly I was in trouble. There was something about this supposedly familiar road that I had not noticed before. Out in that water slick I could see just the very tops of the roofs of a couple of cars. This wasn't just a wet road, there was a deep dip in the road that had created a pond which had swallowed the previously unwary.

Too late. The front of my VW started to lift off the pavement. I quickly put the car in reverse but it didn't do any good. I no longer had enough traction. With the car's forward momentum I drifted out to sea!

Think quick! I could feel a blast of hot moist air coming up from heater vents. Water was starting to rush in through those ducts. I jammed the levers down sealing off that source of leaks. My car simply sat there, bobbing merrily without any sign of sinking.

The motor was running fine and the car was already in reverse so I very gently pressed the accelerator. Best not to cause too much of a stir. The slowly turning rear wheels started to provide some propulsion. Slowly, very slowly, the car drifted back until the rear wheels touched solid ground. With traction I could now pull the car out of its predicament.

After that I drove the long way around the Rose Bowl area and had a different hair-raising adventure with my friend and the rain, later in the day.

True story! No tall tale. VW's really could float. The only penalty for my foolishness is that I had to have the wheel bearings repacked, and that was expensive.

Location! That being the measure of value in real estate. By that standard U.C. Irvine had the world's greatest location for a campus since it was the inland extension of Newport Beach and its Harbor. Newport Beach — one of the premium garden spots of the planet. Golden rolling hills cascade down to envelope a magnificent harbor that once was a premier trading port for the sailing ships of old.

Newport Beach, whose salubrious climate is matched to sun, sea, surfing, sailing — and wealth. As an added attraction, nearby Los Angeles is one of the world's greatest cultural centers. In those days, too, the immediate area was the center of aviation, the space program and electronics. It was the technology capital of the nation. It was no wonder that even before the establishment of the university great gobs of money had poured into Newport Beach and later into the rapidly developing city of Irvine.

While I was a student at the university I became the protégé of a technology investment angel. My very smart boss had the knack of finding gold in unexpected places. He used Newport Beach money to buy minority investments in struggling companies — never more than 15% of the company was his rule. He then would bring in technical and financial experts to diagnose and rejuvenate those companies. When the companies were again thriving he would sell his interest and his investors would make a handsome profit. It was a win for all: the investors, the companies and my boss.

This being car culture what better way for my boss, and his two partners in the enterprise, to celebrate their new fortune than with a new car. After a bit of searching they found the perfect vehicle. Suitably eccentric, it was completely in their character. It was a very long pre-war limo that had belonged to the Queen of Belgium. It had all the fixtures that you would expect of a ceremonial vehicle: external running boards for the guards to stand on and vertical handles for them to grasp, gilded fixtures, all. My boss bought fancy chauffer uniforms for himself and his comrades and the trio of them took turns driving each other around town as if they were royalty.

## THE TREK CONTINUES

It was from my boss that I learned just how much wealth had concentrated in Newport Beach. He remarked one day that there were a hundred thousand bank accounts in that town with more than a million (1960s) dollars in each. Allowing for inflation, that represents trillions of liquid dollars in today's money.

Sadly, in 1969 the capital gains tax was raised, and within a year, technology investment money dried up. For most of the 1970s, until the tax was again lowered, it was a bad time for risk investment. The excess liquidity went mostly into safe, inflation-proof real estate, driving up prices and thereby making housing unaffordable for many. Unintended consequences, you see. My boss, who just happened to have a PhD in physics, and had once taught at Cal Tech, closed down his operation and went off to the Scripps Institute to do cancer research. Without his steady stipend to pay the bills I dropped out of school and got a full-time job.

During my occasional sojourns visiting the wealth of Newport Beach, then and in the years thereafter, I was introduced to some distinguished members of the community. For me, the most interesting thing about these people was how down to earth they were. I had the feeling that they were still a bit surprised at their prosperity. Most had earned it, of course, through hard work and intelligent risk taking. Although they enjoyed their new toys, none I met were stuffed shirts. These people were wholly unlike the arrogant Tech Lords of today. Rather, they were good-hearted souls who enthusiastically gave their time to charitable organizations — particularly those assisting underprivileged and handicapped children.

As a student I availed myself of the university's excellent program in sailing — Newport Harbor being the perfect place to learn that skill. Lots of fun, and I did become a proficient skipper. But my real outdoors recreation there was surfing. And boy did I surf!

Between the Trestles at San Onofre in the south and the pier at Huntington Beach in the north are located some of the world's premier surfing spots. Indeed, Huntington Beach — Surf City as it is known — holds the world class U.S. Surfing Championship each year in September. That is the time when, like clockwork every year, machine-perfect waves march up from the south.

In the early summer the water on this stretch of coast warms enough that a wetsuit is not really necessary for the hardy. Early each morning, before the wind picked up and blew out the waves, I would head to one or another of my favorite spots for a couple of hours in the water.

One morning I decided to see what was happening at the Santa Ana River jetty. It wasn't a great surfing spot but it could be a fun place to ride with a belly board. I grabbed my small board and shoved it into the back of my VW and headed down to the jetty.

The river's exit into the ocean was at the far end of West Oceanfront Drive, at the end of the bluffs. In those days the beach at this west end of town was relatively unused by the public. Part of the problem then was a lack of street parking. This didn't matter to me since my bug could park on the beach sand without getting bogged down. My car was, after all, a latter day Kubel Wagen. When I arrived three or four surfer VWs were already parked on the beach. I pulled in next to them and started to walk down to the surf.

Just then a Cadillac pulled up and prepared to also park on the sand. A Cadillac is a wonderfully comfortable car to ride in but it had no business being there. I backtracked and flagged the fellow down, warning him his car was going to be trapped.

It was clear this man was not a local for he was dressed in a dark suit and tie instead of the casual dress that local businessmen had already adopted. He wasn't going to take advice from some surfer dude so he drove onto the sand anyway. Bam! One Caddy, beached. I just shook my head at the manifest stupidity. This guy was going to have to walk back nearly a mile to get help and then he was going to have a hefty toll to pay to get his car back on the pavement.

After a couple of hours I came back from playing in the waves. The Cadillac was still there waiting for a tow truck. The now bedraggled gentleman was sitting disconsolately in his car, door wide open.

I got in my bug and carefully drove past the man and his car at slow speed so as not to kick up any sand. I was polite. No need to cause more misery for the stranded traveler. Still, I do remember the look on his face as I drove back onto the pavement and then away.

## BUG AT THE BEACH

*Just like Chester's mid-70's VW. Now, lo these many years later, both are classics. —editor's note*

# THIEVES

Sarah and I had decided to see a movie. The movie was too controversial to play at our neighborhood theaters. So we were compelled to drive down to one of the small art theaters that abound in West LA. Sarah met me at my office, a large rococo building perched high above a nearby freeway. She left her car in the basement parking lot and we drove off to a local restaurant. After dinner we returned to the parking structure and exchanged cars, Sarah decided she wanted to drive, so we left my car in the hidden and safe confines of the building's basement.

The movie was a stinker. All the way back to pick up my car we laughed about its ultra serious tone set amidst implausibility and plain old bad acting. When we arrived back at the parking structure my car was nowhere to be seen.

With mounting fury we drove to the police station to make our report. The policeman on duty looked at us with bored disdain — clearly it was our own damn fault for driving an automobile that was currently popular with car thieves. Though I remained polite, inside I was furious at his superciliousness. How was I to know that a two year old car could be a theft magnet.

While I was seething, the policeman explained the mechanics of car theft. Thieves troll for cars such as mine. They wander about the streets until they catch sight of a potential victim. They then follow their quarry until the car is parked. If the car remains accessible after the driver leaves the thieves pounce. In only a few seconds the stolen automobile is long gone. We had no more chance of our car's recovery than we had of taking a vacation on the moon.

The next few days were taken up with the usual insurance issues. We were given a few days of grace to allow the police to find our car, but the insurance company advised us, again, that the situation was hopeless. And so it proved to be. I was shocked at how little the company was willing to allow on a still relatively new car. Their offering was well below Blue Book. But we had no choice. We had to accept their offer.

And so, Sarah and I were on our way to buy a new car. She had already decided on what it should be. Not long before, a new sports coupe had been introduced on the market, the Acura Integra, and it had already developed an outstanding reputation. Sarah loved sports cars and she insisted that we go for a test drive. I was reluctant at the time because the payments would be higher than we had been paying, but as the old song goes, "Whatever Sarah wants, Sarah gets."

The car salesman was most enthusiastic about this model. A tall, lanky man, he nevertheless was somehow able to shoehorn himself into the rather cramped rear seat, and we were off. I drove the car first. Winding up and around the hills, I was very impressed with the car's performance and handling. After a bit Sarah insisted that it was her turn. We exchanged places and were off once again.

One thing the salesman did not know about Sarah. The way to survive the experience of riding with her at the wheel was to say a prayer, strive for the tranquility of Buddha, and leave your fate in the hands of God. Though I had driven the car hard around those hills, Sarah practically left them on fire. It's not that she was a dangerous driver, she wasn't, but her skills and habits, with her experience ranging from driving school buses to racing cars, were far beyond the norm. I once knew a famous astronaut, probably the best jet pilot of them all, who had a similar reputation behind the wheel — and people were extremely reluctant to ride with him as well.

"Nice car," she said when we arrived back at the show room, tires squealing. "Kinda reminds me of my old Porsche." As I got out of the car I glanced back at the salesman. He was frozen in place, his hands grasping the convenience handles so tightly that the white bones of his knuckles seemed almost to burst through the skin. He finally emerged from the back seat, dripping with sweat, and proceeded to stagger a wavy line back into the building. The good thing is that he was in no condition to drive a hard bargain.

Some days after we had taken delivery of our new car we got a surprise phone call. It was the District Attorney of Santa Barbara County. They had our car and they wanted us to drive up, identify it, and then give court testimony. The car was in Santa Maria, still further up the coast. It was a mess: half disassembled, the front end gone, the hood gone, the dash panel in pieces, the seats dismounted and thrown back into the cabin helter skelter. Poor baby.

Later, at the testimony, the District Attorney told us what had happened. First, he thanked us. "With your help, we are finally going to able to nail those bastards. We've known who they were, but we've never been able to pin anything on them. Now we can, *big time*. They are going off

# THIEVES

on a very long, and rather unpleasant, vacation." Then, he told us the story.

It was a highway patrolman who caught the thieves — one of them, anyway. The rest were rounded up shortly thereafter. It seems odd that the bandits would make their home in Santa Maria — that town is the home base of a major center of the Highway Patrol. Being there sounds risky, but the thieves may have thought "under their noses," so to speak, was the safest place to be.

At any rate, this patrolman was waiting near an intersection. He was off duty, and this wasn't his jurisdiction, but still, he was alert. In front of him a rather derelict car ran a red light. The patrolman pulled the vehicle over and checked the license plate. Sure enough, the license came from a stolen car of the same make and model. Only the license wasn't from *our* stolen car. And yet, the body serial number later confirmed that the car was indeed ours.

The thieves had been busy disassembling our car for its parts when they discovered that their hoist wasn't high enough to lift the engine free. They decided to move the car to another location with a taller hoist. So, they dumped the front seats back in the car without bothering to bolt them down, slapped a handy license plate on the car's back, and sent the driver on his way. Well, the car was far enough gone that a panic stop for a sudden red light they encountered would have been suicidal, so the driver just kept on going. There is an urban legend that thieves are invariably smart and wily. Not so. This bunch proved to be a collection of bumblers.

"Why did they want to strip the car, rather than take it down to Mexico, or someplace, and sell it off?" I asked the D.A. His response: the car was worth *far* more as a bag of parts than as a complete vehicle. "Go to any auto parts store and price out used replacement parts. You'll see. What's left of your car now belongs to the insurance company. They'll make a ton of money on it. At your expense, of course."

# POISON

LET ME TELL YOU A TRUE STORY. There once was an evil wizard who had a propensity to poison people — poison them with a magic potion so deadly that, if they took it, they were sure to die almost immediately. Really, the wizard didn't consider himself evil. Poisoning people with the deadly potion was just one of the things he did for a living. He actually thought of himself as a very nice person.

One day a handsome prince was not feeling very well so he betook himself to see the court physician. The physician poked and prodded the prince. He even leeched blood from the prince and sent it to the court alchemist to discover what humors it might contain.

Sadly, the court physician informed the prince that he was dying, that he could drop stone cold dead at any moment. Aghast at the news, the prince begged the physician to do something, anything. Perform a miracle and the physician would be handsomely rewarded by the kingdom's treasury.

The physician was honest. Alas, he could do nothing, he said. Nothing himself, that is. However, there remained one chance, one chance only, for the handsome prince. There were dark rumors that the evil wizard, through his mystical arts, might extend the life of the prince through mysterious alchemical magic. But there would be a price to pay. The handsome prince did not ask the price for any price would be

## THE TREK CONTINUES

worth continued life. So the honest physician wrote on some parchment and sent the prince off to give the note to the evil wizard.

The evil wizard told the prince that, for substantial gold from the kingdom's treasury, he might be able to save the prince's life. The prince would have to take a magic potion to survive. The evil wizard, in a fit of truthfulness, told the prince that this potion was very dangerous, but it might help. The prince ignored the dangers. He wanted the magic potion at any risk.

The prince took the poison potion. Within seconds, the potion began to take effect. Only then did the prince realize just how dangerous and lethal the deadly potion really was. But it was too late. The potion was already killing the prince. The evil wizard then revealed the truth. The prince was now in thrall to the wizard. Each day the prince must take an antidote to the potion or he would die.

The evil wizard couldn't be bothered administering the antidote every day himself, so he turned the prince over to the court alchemist to get the daily antidote. The good news is that the prince *did* survive the deadly disease that had sent him to the court physician in the first place. The bad news is that the prince was now perpetually chained to the daily antidote and therefore to the evil wizard. The handsome prince lives on to this very day — every day forced to take the antidote to the evil wizard's deadly potion.

The handsome prince is me, of course. Well, I don't know if the face I see in the mirror is handsome or not, but let's pretend. I really did take the wizard's deadly poison. It really did start to kill me within seconds. I really do, every morning, have to take an antidote to the poison to keep me alive. The alchemist is my local pharmacist who keeps me supplied with the antidote. As far as the wizard goes, I see him every year and he subjects me to probes by various machines which look inside me to see if I am still alive. It seems that I must subject myself to his will for the rest of my life so I really am in thrall to him. And, the wizard really does think of himself as a good guy.

This tale of alchemy and my ongoing bouts with the evil wizard began when I was forced into retirement a few months before my seventieth birthday. Retirement meant that I finally had the time for orthoscopic surgery to correct a decades old nuisance — a damaged knee. The knee had been injured during a hundred mile backpack trek in the depths of the Grand Canyon.

I flunked my pre-surgical physical. The anesthesia would likely kill me, I was told. I had developed a very serious hyperthyroid condition which caused an uncontrollably rapid heart rate. My heart could, at any moment, just stop. And my thyroid was so out of control I was just short of having a deadly full blown thyroid storm. That, too, could kill me unless I got to emergency immediately: my thyroid glands had to be removed, STAT. That was the bad news. The good news was it was fortunate I retired when I did so that this problem could be discovered. If I had stayed with the usual stress of the job, as I originally intended to do, I would not have seen my seventieth birthday.

Since thyroid surgery was out of the question I was stuck with the alternative: I must take the poison pill. The poison is Iodine 131, a highly radioactive substance. It emits beta and gamma rays. Iodine 131 is a major, and most dangerous, fallout product from nuclear explosions. Because it is iodine it is immediately taken up by the thyroid gland. Because it is highly radioactive, a massive dose kills the thyroid in a few months. Paradoxically, in very small doses, the radiation causes fatal thyroid cancer.

In times of peace Iodine 131 is made in a nuclear reactor by the transmutation of thorium into iodine. Because of its short half-life of only eight days it is highly radioactive. The short half-life also means an alchemist must create the dose immediately before it is to be administered. Thus, the poison pill is manufactured and administered on a very tight schedule.

On the specified morning, at the specified time, the Geiger counter that I had brought along with me started clicking away as I walked down the long hall towards the radiology clinic. There was something *very* radioactive within that suite. Unlike the usual leisurely doctor's office visits, this time there was no wait. As soon as I walked through the door I was hustled into a shielded room where a lead aproned radiation technician joined me immediately. Already in place in the room was a large cylinder. Even capped with a massive lid on the thick-walled lead cask my Geiger counter was going crazy. Whatever was inside that thing was *hot!*

The technician handed me a small cup of water. He lifted the lid of the cylinder and reached deep inside with long tongs. Out came an ordinary seeming gelatin capsule which he dropped into my hand. My Geiger counter screamed into saturation, its digital dial was spinning up so fast you couldn't read it. The technician immediately backed away to the far side of the room and started fiddling with a scintillator. For a few seconds I just looked in wonder at the noxious thing I had been given. I was holding death in my hand. But I had made the commitment. Shrugging my shoulders, I put the capsule in my mouth and swallowed it. Now, *I* was far more dangerous than that sealed lead container had been. The technician took a quick reading with his scintillator, nodded his head, and *fled* the room.

No one stopped me when I left. There was no paperwork to fill out, or even sign. Everyone quickly backed away from me. I had never been so shunned. All were visibly glad to see me go. Before I reached the parking lot my Geiger counter told me that the iodine had already migrated from my stomach to my thyroid. Boy, that was quick! Within mere seconds my thyroid had come under lethal attack. It wouldn't survive long. Without the antidote I would soon die.

Time to hide out. Now dangerously radioactive, I was a hazard to others. I was especially dangerous to pregnant ladies and small children. So, back to my cave I must go, and stay. For months to come I would live the life of a hermit.

Actually the quarantine isolation wasn't so bad. There was plenty to do. I resumed projects that I had put aside years before when the demands of earning a living had gotten in the way. To keep me company I had a good library and a fast internet connection — great time consumers, both. And, I had two cats. Besides, I had a phone so I could talk to friends even if I could not be with them. Time passed.

My cats were my biggest problem. They wanted to cuddle with me. But I couldn't let them — I was too dangerous. I kept pushing them away. This they would not tolerate. They got *very* upset with me. In the end I partially gave in. Better to occasionally give them love even if there was a chance that I might be hurting them. This proved the right thing to do.

How radioactive was I? As a test I put my Geiger counter in the kitchen and walked to another part of the house. There were two walls and a couple dozen feet between me and the Geiger counter. The loud buzz from the instrument said I definitely was unsafe. In terms of gamma rays I most assuredly glowed in the dark.

If I needed to push my cats away to save them from harm, what about me? After all I had a nuclear weapon blazing away a mere four inches below my brain. Some injury to that most vital organ — the essence of me — must be inevitable. Indeed there was loss, but the deficits were slight and subjectively almost unnoticeable. The best description I can give is that some signs of aging showed up before they normally would. Mostly these involved a certain frustrating slowness in recalling names and words.

Before my rendezvous with the poison pill I had overstuffed my freezer and acquired plenty of thyroid hormone pills — the antidote. There was enough food to keep me out of circulation for weeks. But in time my refrigerator was empty. I must break my quarantine to restock. Since the supermarket was open at all hours I figured it was safe to go there very early in the morning while the store was mostly empty. So I set my alarm for an ungodly early hour and dragged myself out of bed. Off I went to do my shopping.

The sun was just starting to rise when I entered the supermarket. The place appeared almost deserted. The only person I could see was the lady at the single active checkout counter. I started my usual wander around the market. I was part way up one aisle when turning into the far end came a visibly pregnant young woman accompanied by a pair of munchkins. I immediately turned tail and dashed back down the aisle.

## POISON

Choosing another aisle, one as far as possible from the previous, I proceeded to start pulling items from its shelves. Once again, approaching from the far end of my aisle was a visibly pregnant lady — a different one this time, and alone. Again I fled and searched for an empty part of the store. Upon encountering a third pregnant woman, this one with a single child trailing along, I knew the situation was hopeless.

Of course I continued trying, as best I could, to avoid these women. The result is that for more than an hour pregnant ladies chased me all around the market. I found myself quickly dashing in and out of aisles as the opportunity presented itself. Those women must have thought that lunatic me had escaped from some asylum.

Eventually my basket was full and I proceeded to the checkout. What, I asked, were all these pregnant women doing here at this unspeakably early hour? I mentioned that I had a medical condition that required me to avoid getting too close to pregnant women. The checkout lady laughed. She sees different ones the first thing every morning. This is the most desirable time for pregnant women to shop.

The solution to my shopping problem, I was told, was home delivery. I hadn't known about that. For a few months home delivery kept my refrigerator well stocked. Problem solved. All I had to do now was wait things out.

After about two months the radiation having washed the thyroid away, I could once again be human. Even better, I no longer had to worry about dropping dead at any moment. I escaped confinement and returned to civilization. It was good to be alive. It was good to be free. It was good to have this adventure behind me. And, it was good to have another story to tell.

*Woolsey Fire photo taken by the author as smoke billows over the top of the mountain, heading toward his home*

# THE WOOLSEY FIRE

NOTES FOR MY NEIGHBORS AND FRIENDS

## SATURDAY: NOVEMBER 10, 2018

IT WAS A DARK AND STORMY NIGHT. But I'm okay. You're okay. Really, I am, okay. So is the house, except for some minor wind damage and smokiness.

## THURSDAY NIGHT, NOVEMBER 8, 2018

IT REALLY WAS A DARK AND STORMY NIGHT — extra dark with the threat of hovering smoke clouds and gale force winds. At eleven-thirty p.m. the phone rang. It was a robocall from the sheriff warning: *Get out! Mandatory evacuation.* What? The Thousand Oaks *Hill Fire* was burning across the city and the wind was blowing it away from me. That did not threaten me. Fifteen minutes later the phone rang again. Then fifteen minutes after that. Persistent devil! I went on the web and checked emergency. Sure enough the area from the far east of me and right up to my immediate neighborhood was to evacuate immediately. I checked the news channel and discovered that a *different* fire, the *Woolsey Fire*, had started some fifteen miles to the east and that was the one that was coming my way.

## THE TREK CONTINUES

And, since I could now see smoke clouds, I should pay attention. I quickly threw stuff into a suitcase and gathered some papers and sentimental things. And, my backup hard drive. Into the car they went. What to do about my cats? Plan A: I would have to find them. Not easy. With things so far out of the ordinary they were already well hidden. Even if I could find them they have not been in a car since they were kittens, fourteen years ago. I was not sure I could handle their panic. On the other hand, I live on a hill and the back area is not only mostly clear of vegetation (it meets city fire code) but there is a clear run down the streets below to safety. So, Plan B, I would open the back doors and let the cats run free. They would be safe out back.

I live in Thousand Oaks, California. My home is on the side of a hill with wilderness to the east of me. I picked this location because, from having lived through the devastating Glendale fire back in the 60's, I believed the layout to be relatively safe from fire. In the years after we moved in here, thirty years ago, the county would periodically burn off the hillsides, and clear the fire break, to prevent even worse. Later they stopped. Decades of tender growth have made the surrounds a magnet for flames. Still, I believed the house was safe. But early hours into Friday I definitely had doubts about that.

Periodically, after the evacuation warnings, I would go out and check the sky to the east. Eventually it started glowing orange and red with reflections on the smoke clouds. About one in the morning I smelled smoke and went outside. The fire had crested the hill to the east. A massive inferno with flames shot hundreds of feet into the air. A whirling horizontal tornado of evil golden fire was grinning right at me and racing

## THE WOOLSEY FIRE

to engulf me. Scary! Really frightening. Anyone who has been downwind of a true conflagration will surely be a believer in fire breathing dragons --- especially the golden ones. They are real! Dragons aren't cute. They are terrifying!

Time to skedaddle. Try Plan A: Find the cats, get them into their carriers and out to the car. No luck. Sparks and Merlin have disappeared, as expected. Out of time, *danger is near!* Therefore, Plan B: Give up. Open the doors to the back. Quick, get in the car, drive the hell away. Hope the cats run to safe ground in the back.

By this time glowing firebrands were skating up the street, driven by the howling wind. The smoke was opaque. I felt my way through the worst of it, drove down the hill a ways and joined the caravan of cars fleeing the area. At the bottom of the hill a long line of cars waited patiently for the signal to turn green at the intersection, the nearby hill a blazing inferno behind. Follow the law even if it makes no sense!

A mass of cars awaited me at the evacuation center. I finally found a parking spot way off in the distance. A long line of people waited silently outside the center. I joined the line. It didn't move. The line got longer. It still didn't move. After a while I decided to scout the situation. The lady at the door said that the line was to register for one of the cots that had been set up inside. I let her know I didn't need a cot. "Go right in," she said. I went back and spread the word, and then went into the building.

People were clustered around a TV, watching, hoping against hope. Cots were laid out on a basketball court. Tired people were sleeping. I sat with a neighbor who had evacuated on the first warning. I brought her up to date with what I had observed. Her house was closer to the conflagration than mine and it didn't look good for either of us. I tried to console her, and myself. By that stage I had pretty much given up hope. What I had experienced was demoralizing. I started speculating about the future. Rebuild or relocate? My wife being deceased, and me being retired and probably no longer burdened with possessions, I was going to be free to do what I pleased.

I walked to another nearby venue which had recently opened up. Outside, a TV news crew was setting up shop. Fire trucks had finally arrived. I was a bit peeved that it seemed that the firemen had taken hours to get to my neighborhood. Someone said they had been busy. Indeed! (I was wrong. Days later I discovered that firemen had indeed been on my street and had worked to save the area.)

In the new place a large screen TV overlooked long tables in the auditorium. Refreshments were plentiful. The hundreds of people around me were visibly exhausted, worried and demoralized. Once they laughed quietly at a morbidly bad joke on the TV.

It was not long after midnight, when I had fled the house very early that Friday morning. A few hours later, about four-thirty a.m., I went outside the evacuation center. The fire glow on my hill had diminished and moved away from my area. Time to go and see about my future. As I drove towards home cars were stopped at the same intersection that had delayed my neighbors and me earlier. Flames were now sweeping the hillside across the street. People were transfixed by the drama and didn't want to proceed. Eventually, cars started to move and I was able to break free and head for home.

And home was still there, the house brightly lit and welcoming. Miracle! Miracle too, for almost all on my street were safe, as were most in my immediate neighborhood. The great losses were plenty, but they were a couple of miles away.

And, the cats, who ran to the front door when they heard me drive up, were overjoyed to see me. What a relief to see them! The cats were all over me, as we sat for a while comforting each other.

Upon inspection I found that part of my fence had blown away and there were piles of ashes and debris in the house that had blown in through the open doors, but otherwise all was undisturbed. A couple of hours cuddling with Merlin and Sparks and taking care of essential repairs and chores, and it was already dawn. I did get a short nap until nine in the morning when my cleaning lady was about to arrive. Chaos always came with that good woman. No more sleep for me. Thirty-six hours with only a couple of short, interrupted naps. Oh well, plenty of time to rest over the weekend.

All's well that ends well, they say. Just another adventure for me. But not for the sixteen hundred families who lost their homes, and tragically, the three people who lost their lives in the Woolsey Fire.

## Sunday: November 19

THIS MORNING I WALKED UP to the end of the street to survey the hill from a different angle. One house there had burned down but that was the only one. At the street's end there was a hotshot crew preparing to walk into the wilderness. Their objective, they said, was to search for smoldering hot spots and put them down. I thanked them and asked that they pass along all our thanks to the others working this fire.

## THE TREK CONTINUES

Surveying the hillside it was obvious why this street had been saved. The homeowners across the street had themselves cleared the firebreak. Since the firebreak extended up the hillside a considerable distance the fire was simply not able to get down into the wooded areas and the homes themselves. Fire goes up, but only down if the winds carry that way. Generally, in a firestorm the winds go up the side of a hill because of rising hot air, and only very rarely sweep down.

Later in the day I met one of my neighbors from across the street. He and another one of our neighbors had stayed to fight the fire with garden hoses. They were not about to let their homes go without a fight. In combination with their brush clearance those two may have saved the whole street. Those of us on living here owe them both a great deal of thanks.

Friday I continued to watch fires burning in spots on most of the hills around me. That lasted the whole day. One particular burn was behind some houses down the hill from me. Police were there but they didn't see the need to call in a fire truck. I was visited later by those homeowners. They wanted to know if I was okay. Very kind of them. I had watched the flames approach their house but none of the trees or bushes caught fire and the smoke died away. They told me that the flames crossed into their backyard and burned some grass. The fire reached within two feet of the house before dying out. As an example of the difficulties that the evacuation caused so very many people, these good people finally found lodging in Claremont, some eighty miles away. Everything nearer was booked solid.

Later my next-door neighbor told me that he and some others had been working the fire behind the downhill home. I didn't see them because the trees between them and me were too thick. But that did explain why the police did not seem worried about the flames. And, very soon the problem there was licked.

The burn on the hill exposed many small caves. I take them to be coyote dens — off and on. The hotshot crew said that not all coyotes survived in this area. They had found one carcass. I suspect we have lots of burned critters. Starvation will be the next problem for predators around here. Probably the more mobile of them will wander off to the north, away from the burned area. Still, the burned area is substantially more than 100 square miles. We won't again see red tails and falcons for some time to come. Fortunately the breeding season is over so no chicks will be lost.

Yesterday and today I have been monitoring the VHF fire dispatch radio. As of mid-afternoon things are still boiling. Some areas, to the northeast of here, that were supposedly completely burned over, are again active with flareups. Even though the winds are still strong and gusty we now have a great deal of air support — the guys on the ground are constantly calling for it. Also, apparently the fire departments are experimenting with drones to see if they can help spot and evaluate fires. That is very interesting.

## Several Days After The Fire The Doorbell Rang

At the door was a tough, grizzled fireman. He was just checking to see if all was okay here.

Little Sparks snuck his nose around my legs. The grizzled fireman melted. Even the toughest is charmed by such a sweet kitty.

THE TREK CONTINUES

Down at the base camp, he said, firemen would come off the fire line, perhaps after having lost a house, and would be found on the ground snuggled up with a rescue dog, recouping morale for the next attack.

"Dogs are good," he said. "Cats are just as good," he said.

*Sparks reading about System Architecture*

# FRIENDS AND FIENDS

When Sarah and I first got together I was living in Orange County, California, picking up various consulting assignments. Then, shortly after we were married, a remarkable opportunity opened up and I accepted a job a hundred miles to the north. This was to work at a company which was the principal technical assistant to President Reagan's "Star Wars" program. One of this office's specialties was rocket engines. There I was surrounded by the royalty of the rocket engine community — a fascinating bunch, those.

Not long after I joined, the company held one of its occasional family dinners. It was there that Sarah was able to get to know some of the remarkable characters that I had been telling her about. Vivacious Sarah was an instant hit of the party and everyone gathered around her to soak up some of her sunshine.

These company dinners began with people standing around chatting. Then Sarah pulled up a chair. Her table instantly filled up. The other tables gathered people in the order of proximity to Sarah's. The far table belonged to our boss, the very peculiar Jason. You might become acquainted with Peculiar Jason in the story "Psychopaths" in my first book. Peculiar Jason's table gathered in the few leftover stragglers and remained mostly empty.

# THE TREK CONTINUES

Now anyone who has observed organizational dynamics knows that it is the table of the Boss that invariably fills up first. People naturally want to pay their respects. Not in this organization. It was clear to all that the real center of attention in those gatherings was my dear Sarah.

Sarah instantly became friends with all, except Peculiar Jason. Sometimes I was seated so that I could glance in his direction and his table of silence. I don't think it was my imagination that Peculiar Jason seemed focused on Sarah, apparently brooding dark thoughts about the lively laughter at our table. That certainly did me no good during the following days. In the evening after a long day's work, I could keep Sarah up-to-date on the events at the office. Sarah, knowing one and all, could laugh with me, cry with me, and counsel me on how to deal with Peculiar Jason.

Later, after Jason had booted me out onto the street, Sarah no longer had the opportunity to meet my new professional friends. The distances between home and the various places of work were just too great for socializing. And of course there were those, the good and the occasional bad, who had come before I met Sarah. For these Sarah had to rely on my stories about these very interesting people. And they were well worth becoming acquainted with, as you shall see.

# WAR STORY

THIS IS A WAR STORY. It really is, though there are no bombs exploding, nor bullets flying. But then, not all battles in a real war are fought with cracking rifles and blasting bombs. Nonetheless, lives, many lives, depended on one man's courage.

The hero of the story was my boss, Don Shafer. The arena for his small battle in a long war was his Advance Design Department, where I worked after receiving my Bachelor's Degree. This was at Hycon Manufacturing Company in Monrovia, California, the company that I lucked into because I did everything wrong in my job search (See "Failure, Part 1").

You may remember my mentioning in that story Trevor Gardner, the founder and president of Hycon. He had returned to the company after serving as Assistant Secretary of the Air Force under President Eisenhower. This little Welsh dynamo had apparently been a holy terror up and down the halls of the Pentagon. Publicly he was the pioneer, champion and boss of the Air Force's ultimately successful Intercontinental Ballistic Missile program. Behind the scenes he helped initiate a number of black programs, including the famous U2 aircraft. He resigned his position in 1956 in a highly controversial protest over *his* boss's decision to cut the funding of ballistic missile development. A year later Sputnik went up and Gardner, retroactively, became something of a national hero, his portrait on the cover of *Time Magazine*.

I only met Trevor once, but the memory and images have stayed with me throughout a long lifetime. Hycon had a long hallway near my work area. I came around the corner of this hall in time to see, at its far end, a small group of well-dressed men. Among them was Don Shafer. In the center of the group was this small man whose face I recognized from the cover of *Time Magazine.* As we approached each other, Don halted the group and introduced me to his boss, Mr. Gardner. After a quick handshake I expected the entourage to continue past me. It didn't happen that way, though. Rather, as the others in the group began passing by, Trevor remained and started to chat with me. He asked me what I was working on and what my plans for the future were. We probably did not talk long — just a few minutes, but the interest that Trevor took in this very young engineer has remained an inspiration and a model ever since. A couple of weeks later Trevor Gardner was dead — still in his forties. Cancer, someone at Hycon said, heart attack the official history says, both probably. If cancer, that brief conversation must have meant exquisite physical pain for the man on his last visit to the company he had created and loved.

The Advance Design Department was something of a hobby for Trevor Gardner — to be sure, a serious hobby. His intent in setting up the Department was to gather together the most creative people he could find and then turn them loose. He had another motive as well, which I will come to. While Don Shafer was the manager, the real focus of activity was the desk of Bert Van Breeman. Bert, a jolly man of great talent, was the Project Engineer of the Department. Bert is also featured in "Failure."

## WAR STORY

We all worked in a bull pen, about a dozen of us, our desks lined up in rows. Communication was instant, much faster than today's email. The Department had a unique mix of inhabitants — people well worth remembering. It was a remarkably creative environment.

Leonard Larks was a rotund optometrist, a man with a ready chuckle. Leonard had grown bored fitting glasses. He wanted to do some real engineering. He did, too, along the way inventing an automatic focus sensor. Leonard's creation, it seems, found its way into the Japanese cameras of that era.

I invented an auto focus sensor, as well, one that was better suited to aerial photography. Many years later I was told that my focus sensor had been employed by generations of aerial cameras. I never did get the dollar and the pin for my badge that went with the patent.

### HYCON PATENTS

*Automatic Focus Sensor*

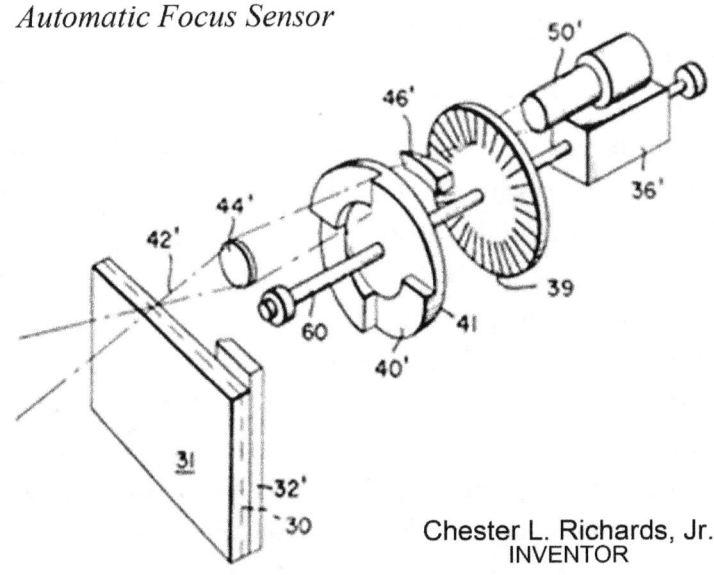

Chester L. Richards, Jr.
INVENTOR

## *Pneumatically Operated Camera Shutter*

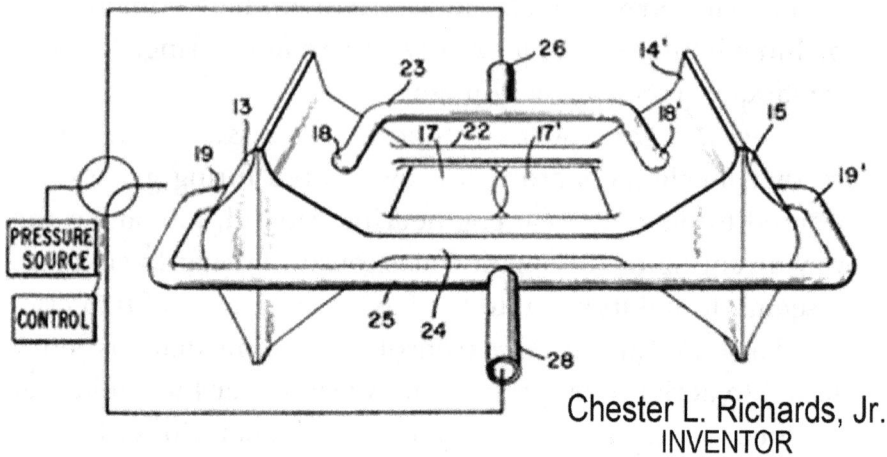

Chester L. Richards, Jr.
INVENTOR

*The Automatic Focus Sensor and Pneumatically Operated Camera Shutter were actually my first two patents. Patent applications were filed in 1966 and assigned to Hycon Manufacturing Corporation. The focus sensor was used in military aerial cameras. The mathematical theory I developed governs focus sensors in consumer digital cameras to this day. The pneumatic shutter was not used in cameras, but it was adapted to control magnetic tape transports for main frame computers.*

OUR ENGLISHMAN, LESLIE P. DUDLEY, a small wiry fellow, was the perfect picture of the British Boffin. Indeed, his wartime achievements in radar and optics had brought him considerable renown. Subsequently he became the premier modern pioneer of stereo photography and display. Stereograms have since degenerated into mere trinkets — advertising gimmicks made of cardboard and plastic. But Les had created the real thing. Les's stuff was magic. Picture a sheet of glass — like an old fashioned glass negative — about the size of an 8x10 photograph. The glass is covered with ultra-fine vertical striations, too fine to be resolved without a magnifier. A casual glance shows nothing special, merely a darkened surface to the glass. But, hold the plate up to the light and something amazing happens. This high precision parallax panoramagram becomes a window into a fully lifelike microcosm. Beyond the window a colorful world filled with living people and all the elements of the real world would be waiting for you to reach through the glass and join in. The eye was completely fooled, so lifelike were the scenes in his windows.

After the Second World War these stereo panoramagrams had had a brief popularity as an advertising medium, in England, and here in the U.S. as well. I remember seeing them, as a child, when in restaurants and taverns. But they proved too expensive and too difficult to mass produce. So now they languish somewhere, mostly forgotten, museum treasures, if any could be found.

FRANK SEGONA WAS A CRUSTY OLD CODGER, maybe in his mid-thirties. He had recently joined the Department after a tour of the coral reefs of the Pacific Isles, installing tsunami detectors, he said. Few in the Department believed him. We had other ideas as to just what he had been installing. Nonetheless, his experience was exotic and interesting.

"What about sharks?" I asked. "They're supposed to be man eaters out there. Weren't you afraid?"

"Nah," he replied, nonchalantly. "We weren't bothered by the sharks. We left the sharks alone and they left us alone."

Then, too, there was a rather pleasant young man, about my age. He shall remain nameless, although I do remember his name. This young man had had a very much misspent youth. But he learned drafting while in prison and Hycon was willing to give him his second chance. He worked out okay.

Another one of the Department's denizens shall also remain nameless, but for a very different reason. Let me tell you a little story to give you the flavor of this particular character.

Tension had been building in the country for several days. President Kennedy was going to Texas. Rumors were floating around — mere speculation it was said — that Kennedy would not survive his sojourn in the hostile heart of Dixie. Instead, the trip turned out to be a triumph. Kennedy was well received by the warm-hearted Texans, on his first visit to the state.

On this particular day, his second visit to Texas, Kennedy was due to arrive in Dallas. We were working quietly when the public address speaker in the office burped awake. "We have just had confirmation," a voice said in quiet tones, "that President Kennedy has been assassinated." I sat there in stunned silence, taking in the shocked expressions of my comrades, my heart gripped tightly as if in a vise. All sat equally stunned.

All, except for one individual. He started laughing gleefully. "I told you all, yes I did," he exclaimed. "I said he was going to get it, and he did. Good riddance to him!" Needless to say, that foul fellow was instantly cast into Coventry.

From then on, only when it was necessary for our work were any words directed his way, and then those words were very few. It became our habit to turn our backs on him whenever he passed by.

Some months later a pair of company guards came into the office and escorted him away. We never saw him again. It transpired that this person had been arrested for espionage. No one shed a tear.

# WAR STORY

The daily routine resumed. One difference — the villain's desk remained empty from that day. The rest of us carried on. At the desk closest to Bert's, sat Will Connell. Will's father, the elder Will Connell, had been an eminent photographer, had left his mark on the movie industry, and had been a cofounder of the highly respected Art Center College of Design. His students, and their students, working as cinematographers, have dominated the Motion Picture, Television and Advertising industries. Will's mom was a firebrand feminist, equally well known in her own right. Will had grown up in the company of famous artists and writers. As a consequence, Will was fearless. A razor tongue, coupled with an equally keen brain, had me trembling with the prospect that he might focus his caustic wit in my direction. It took me a full year of careful observation before I felt confident enough to approach him. Will was educated as a geologist and petrologist. I had a question about growing diamonds. That broke the ice, so to speak. Will and I instantly became great friends. I watched his children get born and grow up. Will, in turn, was Best Man at Sarah's and my wedding.

Perhaps my favorite in the Department was John Brogden — one of our designers. John had not finished high school, but he was able to self-educate to a level where he was hired, and worked for years, as a professor at a noted Liberal Arts college. There he taught both science and art. John Brogden was a giant of a man, both physically and in those interior aspects that make a man a man. To get a mental picture of John, think of a middle aged, but more robust, Gary Cooper. He even had Cooper's shy plainsman mannerisms –- having grown up in the same part of the world. There is a photo of Teddy Roosevelt and his victorious Rough Riders atop San Juan Hill. Standing next to Teddy is a rangy man, Hollywood handsome, a tall, broad brimmed, cowboy hat cocked sideways on his head. That man is the spitting image of John Brogden. It might even have been John's grandfather, who had been the sheriff of a very tough Dakota mining town. So it really could have been.

John had a quiet dignity which came from profound experiences I only gradually became acquainted with.

John, you see, was a war hero — the real thing. Like most true heroes he didn't put on airs — he had just done his job, he said when asked. Indeed, only occasionally would a small story of his military adventures slip out. A tale of a time in the Philippines when he rose from his sick bed, nearly hallucinating with malarial fever, and ran to help ward off a Japanese attack. Or a tale of Okinawa with a massive typhoon bearing down. John and his buddies retreated to a cemetery cave, throwing the skeletons outside to make room, only to be joined in the darkness by enemy soldiers who were equally terrified by Nature's chaos. After the terrible winds had passed the two groups retreated to their respective battle lines and resumed murdering each other.

I didn't fully recognize John's heroism until many years later. This came during a casual conversation with someone I met by chance at a professional conference. He told a story about a platoon sergeant in frigid Korea who had braved a hail of gunfire to carry my new acquaintance to shelter and medical aid before returning to the battle. "He saved my life, John Brogden did. I have never known anyone else like him."

Not everyone I worked with was a member of the Department. One day I was told to go down to the manufacturing area and introduce myself to Gunnar Edenquist. Gunnar was very tall, very slim, silver haired and craggy faced. He was one of the originals in the aviation industry. For several weeks we worked together to solve tricky problems — saving the company many millions of 1960s dollars. That was my introduction to the old style of aviation engineering and to stories about the people who created the industry. Gunnar knew them all.

Gunnar has not acquired the fame of other aviation pioneers, but he was there and he made substantial contributions to the development of aviation. He started his career in the youth of the last century, engineering for Glenn L. Martin. The aviation community was close knit, with people often working together on various projects. So Gunner knew them all — Martin, of course, and Larry Bell, Jack Northrop and the Lockheed brothers. Bill Boeing was a frequent visitor, learning aviation and how to fly from Glenn Martin.

Not long after, Gunnar became Chief Engineer of Kinner Aircraft and designed many of the engines used by aircraft throughout the twenties, thirties and into the forties, engines that are now to be found tucked away in far corners of important aviation museums. Amelia Earhart learned to fly in one of Gunner's airplanes. Ultimately Gunnar became bored with retirement so he returned to the industry as an ordinary engineer. Now he was working at Hycon.

Through my association with Gunnar, I am connected to the earliest pioneering days of aviation. Aviation is still that young.

Last, and certainly not least, was my immediate supervisor, Bert Van Breeman. Bert was a stocky man, built like a tank. This was all muscle for Bert had been a wrestling champion of the U.S. Air Force. To say that Bert was intelligent is decidedly an understatement. For example, he was one of the independent inventors of the integrated circuit. Given what subsequently transpired he may well have been the first to invent this extraordinary technological breakthrough. At the time he came up with the idea, he was working for Honeywell. Big, stolid, self-satisfied Honeywell. They did not see the point in pursuing such an unimportant idea. I would like to believe that Honeywell's management later regretted killing the Golden Goose. Probably, though, those unimaginative executives had long since forgotten about their missed opportunity.

Bert was versatile. If one of his inventions did not catch fire, there was always another, quite different creation, coming along. One such has given a great deal of pleasure to a great many people, for Bert invented something which will be familiar to all sports fans. We've all seen it: the TV starts by showing a panoramic view of the entire football stadium. Then, the camera swings down to the broad expanse of the field below. Swiftly the camera zooms in until we see, close up, the quarterback giving his count in the lineup. We can even read his lips. The 100 to 1 zoom lens, which made such a journey possible, was one of Bert's creations. I won't go into the details about it except to say that it was, conceptually, very clever.

Late each afternoon, when our work had naturally wound down, a small group of us would congregate around Bert's desk. Our gatherings weren't large. Typically there would be anywhere from three to four of us. John Brogden and myself were the regular core of the acolytes. Once we were comfortably seated, someone would suggest a problem, or a need. It could be anything — it did not have to relate to Hycon's business, and usually it didn't. To this day I'm not sure exactly what the magic was at each of those conclaves. What I remember is that invariably, and in short order, creative solutions to the posed problem would bubble up, be tinkered with, refined and, finally, polished into a mature invention. These creations were works of art. The common characteristic of all of them was their simplicity — *elegance*, it is called by engineers — and obviousness. However, these inventions were obvious *only in retrospect*. This is one of those things that make you say to yourself, "Why didn't I think of that?"

Seldom did a day go by when Bert's gathering did not produce at least one patentable and marketable invention. Few of these were of interest to the company, and each of us had other fish to fry, so most of the inventions, no matter how attractive, were abandoned. Nevertheless, the true goal of the daily séance was achieved. We all became highly skilled practitioners of the Art of Invention. And, I must say, only rarely in all the years of my professional experience have I found anything quite as entertaining as Boss Bert's Sunset Séances.

One thing about my work environment was obvious from the beginning. Much of the time the office was mostly empty. Where *was* everybody? Inquiries usually were met with some vague mumblings about Mountain Avenue. It soon became apparent that there was much more to the Advance Design Department than was evident on the surface.

The full import of these absences did not become apparent for nearly a year. One morning I arrived at work to find my colleagues clustered around the latest copy of *Aviation Week*, a trade magazine notoriously nicknamed "Aviation Leak" for its ability to acquire highly classified information. The article, in this particular case, provided quantitative details, photos and scaled drawings, expanding on President Lyndon Johnson's revelation, a few nights before, that America was flying an extraordinarily high performance aircraft, the hypersonic SR-71 spy plane. Johnson had felt compelled to announce the aircraft because there were very many reports of UFO's flying above the deserts of Nevada and Utah. There were even signs that panic was starting to set in among some of the people in that region.

No one in the crowd seemed surprised at the existence of the aircraft — only at the details that were being revealed. "They got that right," someone said. "No, that's wrong," was the response to something else. It was immediately evident that my colleagues knew a lot about this aircraft. It was no great trick to finally figure out what was transpiring at our satellite facility over on Mountain Avenue. This also explained the nearby work area with its heavy, combination-locked, industrial green door. It was now obvious to me that the Advance Design Department was primarily a cover for Black World reconnaissance programs.

By 'n by Don Shafer walked into the office. He immediately joined the crowd clustered around the magazine to see what was up. With a start Don grabbed the magazine and began rapidly leafing through the article. His skin went gray as blood drained from his face. Don staggered, on the verge of fainting, with the revelations in his hand. Recovering somewhat, he looked around wildly at the assembly and commanded, in no uncertain terms: "No one is to say anything about this! This information is not to go beyond this room." My colleagues suppressed secret smirks — this particular cat had already clawed its way out of the bag and was long gone.

If Advance Design was a cover for Mountain Avenue, it could even be inferred that the fifteen hundred employee Hycon Manufacturing Corporation was, itself, primarily a cover for the small Mountain Avenue operation. That speculation was confirmed for me some years after James McDonnell, of McDonnell Aircraft fame, purchased Hycon and moved it to another Southern California location. While working nearby, I discovered that the whole of what was left of Hycon was installed inside a large building within a building. The outer building was merely a shell — a Potemkin Village. It had no other function than to camouflage and shield the inner building. The inner building was a highly secure facility which was totally devoted to Black World programs.

In order to be effective in its true and secret role, the earlier Hycon still had to produce viable, visible products. And so, after a long diversion to set the scene and introduce the characters, we finally come to the story of Don Shafer's heroism.

In case you haven't yet guessed, Hycon mainly produced cameras for military reconnaissance aircraft and other, very highly classified applications. Aerial photography had changed dramatically since the days when my father did that work for the Army Air Corps. In Dad's day an aerial photographer would strap on a harness, he would lean way out the door of a lumbering, obsolete bomber and take photos with a large hand-held camera. Until the new war taught us its harsh lessons not much had changed in the Air Corps since the First World War.

But over in Europe, where the fighting was intense, things were very different. The British were the great innovators of modern aerial photography. They mounted fully automated cameras in stripped down, souped-up Spitfires and Mosquitoes and flew them, at low altitude, into the fiery teeth of anti-aircraft defenses.

## THE TREK CONTINUES

At the beginning of the war, we adopted and modified these British mechanical marvels. Here's what we had to accomplish.

War time aerial cameras are faced with a number of technical problems, which lead to great mechanical complexity. The operational requirements of reconnaissance aircraft would necessitate an extra-large and complicated camera in order to take clear photographs while traveling at the high speeds of an aircraft. And, the camera had to fit in the cramped space of a fighter, so the machinery had be very compact and lightweight. Second, and more important, the ground would be whizzing by. Imagine trying to photograph the side of the road from a car going sixty miles per hour. It ain't easy getting nice crisp pictures of beer cans lying in the gutter! Aerial cameras required special mechanisms to move the film at just the right rate so as to precisely track the fast-moving image during an exposure. And, this film movement had to be adjustable depending on the aircraft's altitude and velocity.

Then too, the film had to come from somewhere and it had to go to somewhere. The film was pulled off of a massive feed spool and passed to an equally big takeup spool. Of course the diameters of the film rolls on these spools changed continuously throughout the photographic mission, one got bigger as the other got smaller, so the rotation rates of these spools also had to change continuously in a coordinated way. Choreographing all these competing activities required machinery of very great complexity. A fine mechanical Swiss watch is a child's toy compared to the tangled intricacy of ultra-high precision, but very robust, gears, cams, springs and Geneva mechanisms that were packed into the cramped space of an automated aerial camera of that era.

## WAR STORY

Things got worse after the Second World War, very much worse. The new wars that developed were different but the need for recon photos was just as great, and the conditions quickly became even more stringent. Jets came in after WW II. Speeds soon almost doubled, then they became three times that of the fastest World War II fighters. Frame rates went up accordingly. If that wasn't bad enough, the photo interpreters soon decided that if the wartime four-and-a-half inch square image format was good, then nine inches square was even better. Eventually the defense establishment was demanding nine inch by eighteen inch photographs at four frames per second. And, the space available was becoming progressively more cramped. The mechanical technology was pushed to the limit, and then far beyond. At some point all this jerry-rigged technology was inevitably going to come crashing down — and crash it did. Unfortunately, the crash came with a Hycon camera design.

This model camera was mounted in the nose of a modified supersonic fighter. Every day, and often at night, over in Viet Nam and in other hostile places, these aircraft would fly into heavily defended territory. The planes would usually get shot up, and many never returned. Brave pilots were risking their lives and some were getting killed, but to no avail. On every mission the story was pretty much the same. As the recce plane approached the target zone the camera would be switched on and the photographic run would commence. When the plane returned to base — *if* it did — the film cassette would be dismounted and rushed to the processing shack. The film would be extracted from the cassette, only to find that the film had broken within a few frames after the camera had been switched on. The mission was a failure. A life had been risked to no avail. It was like flying into air-to-air combat only to discover that nobody had loaded bullets into the guns. People got very cranky.

It wasn't just our Air Force and Navy that were suffering. The Israelis were also using this camera and were also at war. Their pilots were experiencing the same deadly frustration as ours. Finally, our pilots mutinied. They refused to fly any more missions until the problem was fixed. Although mutiny is a capital offense in wartime, no disciplinary action was taken. The commanders understood what was involved and the whole affair was hushed up.

What to do? Our regular engineering staff tried, and tried hard, to fix the problem. The mechanism was redesigned, and then redesigned again. Extra springs were somehow crammed into the limited space to help relieve the tension on the film. Nothing worked. Finally, in desperation, the problem was given to the Advance Design Department — a truly drastic move since the regular engineering Department was definitely at odds with us prima donnas. Seldom seen John Dankowski was pulled back from Mountain Avenue and Tiger Teamed with Frank Segona and others. Dan's analysis showed what was wrong. The problem turned out to be philosophical.

Buried in the heart of all those gears, cams, springs, wheels and what-not was the central controlling mechanism of the film transport. Since the days of Edison a Geneva mechanism had always governed the movement of film through any motion picture camera or projector. A Geneva is an odd looking device which converts continuous rotary motion into intermittent stopped motion. When the motion is stopped a photograph can be taken.

# WAR STORY

The origins of this very ingenious mechanism are lost in the mists of time. What is known is that versions of it were used by watchmakers (thus the association with Switzerland's Geneva) as early as the 17th century. The standard External Geneva is robust and reliable, but it takes up quite a bit of room. In the cramped confines specified for this new aerial camera there simply wasn't enough room available. What next? The solution was found by one of the designers of our regular engineering staff. If a conventional External Geneva was too big, well, let's just use an *Internal* Geneva drive.

An Internal Geneva works much the same as as its external cousin but it is turned inside out and more compact. Unfortunately, it was the Internal Geneva that was at the heart of the problem. With some careful design, and a few extra springs here and there, we got a model which seemed to work okay and the film transport was put into production.

Then came the dreadful news that this new design consistently failed in the bone-jarring world of supersonic combat.

As I said, Dan's analysis found the problem, and what a problem it turned out to be! The clever Internal Geneva was just a bit too clever. As Dan discovered, periodically the mechanism had to instantly snap back to an earlier position. There is only one way it could do this. During part of its cycle it had to go *backwards* in time. To go backwards in time it had to move faster than the speed of light. This was definitely a philosophical problem, nothing material, and I do mean nothing, travels faster than the speed of light. No wonder the machine kept breaking, it had to do the impossible. The conclusion was evident: there was no way this mechanism could be made to work.

The Tiger Team tossed out the entire design, laid down a fresh sheet of paper and started from scratch. What exactly is a film transport supposed to do, they asked? Its functions are few and simple, they realized. Film must be drawn from a supply roll, passed with controlled velocity behind the lens and deposited on a take-up roll. That's all. Simple in principle. Then why do we have to do it the way it's always been done? After all, we are professional inventors, they said. Let's see if we can perform those basic functions with a fundamentally new approach.

Good engineering requires simplicity. Throw away all that mechanical junk. Do the job with electricity. The film rolls were constantly changing in size and rotational speed, so they electrically measured these constant movements with a couple of simple potentiometers. In addition, they installed a couple of small motors to move the film so it tracked the moving image. A few transistors, resistors and capacitors coordinated all this activity. Simple, rugged, compact, inexpensive, and it *worked!*

There you have it: in a matter of hours the team had created a fundamentally new design with just four servo motors, five simple measuring devices and a single circuit card stitching together a handful of transistors to control and coordinate the movements.

Simple, clean, and, as they say, *elegant!*

All that was needed, now, was to build the thing and see how well it worked.

The team showed its design to Don Shafer. He liked it and carried it to the president of the company. All he asked for was money. He was turned down. So, he risked his job and career because he was determined to build it anyway.

## WAR STORY

It's not that Hycon's president was an old cranky crackpot character (that came later, when Mr. James McDonnell bought the company and started running it to his taste). The president, in fact, was a very nice, a very accessible and easy going, person. But he obviously was much too cautious. Perhaps he simply thought he was protecting Hycon's existing product lines. I really have no idea what transpired that day, but Don was apparently told, in no uncertain terms, that he was not to pursue this idea any further. Still, Don recognized that developing this new film transport was absolutely vital. It was vital not only to the survival of Hycon, but, much more importantly, vital to the mission of the recce pilots and to the survival of the troops they were trying to protect in this new war we were fighting.

What to do? Now Don Shafer, a very quiet and reserved man, was also as tenacious and tough as a bulldog. During the Korean War Navy Commander Don Shafer had been the senior officer in charge of munitions aboard a fighting aircraft carrier. There was perhaps no more demanding and dangerous a shipboard job in the Navy than that. So, Don Shafer was no stranger to hazard, and to courage.

Don threw caution to the winds and backed the project in deliberate violation of the president's direct order. Maybe he had enough funds from the Department reserves to finance the operation, or maybe he didn't. Where he found the money needed will forever remain a mystery. Don't ask — no one will ever tell. That he was taking a very big risk almost goes without saying. I presume that Don opened a private accounting book, and started juggling funds from various nooks and crannies.

The Tiger Team was moved behind the Green Door — the one with the combination lock — and turned loose. Within a couple of weeks word filtered out that the new transport was up and running. After a bit of debugging the unit worked perfectly — superbly well — even during its shake, rattle and roll torture on the massive shake table. There things sat, just waiting. The members of the Tiger Team were pulled out of the mysterious room and sent back to their usual assignments. Nothing was said, nothing was done. Weeks went by. Months went by (not too many months, however). Until...

One day the president phoned Don Shafer in a panic. The Israelis had called and laid down an ultimatum: They were coming to town in a week and they expected to see a film transport which worked perfectly — no more film breakage, or else. The or-else involved both pulling Israel's very large contract and lobbying in Washington to put Hycon out of business, a sentiment, by that time, fully shared by both the Air Force and the Navy.

What to do, what to do! Was there anything that Advance Design could do to pull the fat out of the fire? What about that film transport design that Don had shown him? Don hemmed and hawed. This was going to be tough. It was going to require major expenditures since the staff needed to work thirty-six hours a day to build something so revolutionary in such a short time. It didn't matter, was the response from on high, the check is blank, pay them whatever was needed — double the overtime. Triple it, if necessary. (And, of course, Don's accounts suddenly would receive more than enough cash infusion to rebalance the books.) Still, Don emphasized the risk. He made no promises. The president was asking the impossible. But Advance Design would give it its best shot.

## WAR STORY

Every day, sometimes several times a day, the president would call Don and ask if there had been progress. Every time he called, Don mumbled (Don mumbled a lot during that week) that some ground had been won, but it still looked very iffy. They had trouble getting the parts needed. There were lots of bugs in the new design. Things weren't operating smoothly. The team was exhausted and making mistakes, mistakes that took a long time to fix. The president grew more and more nervous. Of course, the Tiger Team was having a very good time of it. The new machine was still humming away, running perfectly behind that Green Door. The rest of us were very much enjoying the show.

The dread day finally came. The formidable Israeli contingent marched in, ready to tear the walls down. The president greeted them with a sense of despair. Don had just informed him that he needed to stall the visitors, some more adjustments were underway. Maybe the contraption would work, and maybe it wouldn't. Finally, word came that Advance Design was ready to receive the distinguished guests. The president, anxious with anticipation, escorted the Israelis down the stairs, through the long corridor and up to the Green Door. Carefully, slowly, Don activated the combination and the visitors entered the sanctum. There, sitting alone on the shake table, was The Device.

I'm not sure what the visitors and, of course, the president expected to see. Maybe they were looking for something very complicated, very intricate. After all, for a full generation such had been the complexity of automated aerial cameras. What they saw instead was almost nothing. Two film rolls, a couple of motors, a couple of simple gadgets, and a dinky little printed circuit board, all mounted on a slab of aluminum.

## THE TREK CONTINUES

The film transport was switched on. Smoothly it accelerated the first frame of film to just the right speed. Smoothly, almost noiselessly, it continued to run, each frame precisely metered, each frame's velocity perfectly controlled (this indicated by the squiggles on the nearby oscilloscope). Smoothly it completed its test run. But, after all, this had been a very benign laboratory demonstration. How would it work in the jarring belly of a jet fighter? The shake table was turned on and the power was cranked up. The film transport became a blur it was shaking so violently. Still, it continued working just fine. The Israelis, taciturn as ever, nodded their approval. The president let out a big sigh of relief. Don said nothing.

The film transport was redesigned for mass production and became a mainstay of our aerial reconnaissance product line. Recon flights in Viet Nam and the Middle East resumed. Good guys were saved. Bad guys were not.

And, Don Shafer, having risked his reputation, his livelihood, his career, and maybe even more to do what needed to be done, carried on.

# THE MOSSBAUER EFFECT

THE WORLD OF PHYSICS HAS A PUBLIC FACE: the compendium of published research and interesting ideas. It also has a realm of little known folklore — tales handed down from master to apprentice. One of these is the *Mossbauer* effect. I don't mean *The* Mossbauer Effect. That is quite famous in the world of physics. And rightly so, too. No, the one I am referring to sheds experimental light on the sociology of Academia.

The principals in this drama have long since passed away. The story is known only to a small number of their now elderly disciples. I am one. This story was told to me during my time working in the laboratory of Professor John R. Pellham at UC Irvine, some time after he had left Caltech. Since he was a key figure in this saga, I got the story from the horse's mouth, so to speak...

AT THE INITIATIVE OF RICHARD FEYNMAN, the imminent physics Nobel Laureate, Rudolf Mossbauer was given an invitation to be a post-doc at The California Institute of Technology. Such an unsolicited invitation was unusual — there must have been lots of rumors swirling around about this talented young man. Mossbauer arrived at the Caltech campus in 1960. He had received his PhD from the Technical University of Munich in 1958 even though he was no longer resident there. Something had caused an estrangement. His 1957 thesis project predicted what became known as The Mossbauer Effect, the famous phenomenon I mentioned earlier. In his dissertation he proposed a certain application of gamma radiation from Iron 57. His professor didn't much like the idea and refused to put his own name at the head of the dissertation and resulting doctorate; in the German system the professor is always given primacy and the student is merely the acolyte and journalist of the great professor's great ideas. I very much doubt the professor was happy about what subsequently transpired.

To understand what we are talking about we need some physics. Imagine a radio transmitter that operates at a very high frequency. I mean a *very, very... very* high frequency. This transmitter emits a gamma ray. The transmitter is the nucleus of the Iron 57 isotope. Every once in a while this type of nucleus gets excited in some way. In response, the iron squirts out a very brief radio signal at a very narrow ultra-high frequency.

Enter Mossbauer. What he realized is that Iron 57 potentially could be used as a super high precision measurement device. He proposed an ingenious way to use said device: Sometimes an instrument that is a good transmitter can also act as a good receiver. Such is the case with Iron 57. Mossbauer figured out a practical way to use this phenomenon to measure almost infinitesimal changes in gravity.

What's the big deal, you ask? Actually it is a very big deal!

More physics: In 1905 an obscure patent clerk, working six days a week, still somehow found time to pursue his hobby of theoretical physics. In those days, long before today's stifling academic control of the literature, anyone who knew the language of a science could publish. This obscure amateur scientist, notorious for being a poor student, and rejected by the academic world, published four papers which made him famous for all of future human history.

The obscure clerk, Albert Einstein, in his four papers proved the still controversial existence of atoms by explaining Brownian motion. He showed that light has the property of particles by explaining the photoelectric effect — this got him a Nobel Prize. He postulated the interchangeable equivalence of mass and energy, leading to the famous equation E=MC2. And, not least, he unified space and time into a single geometric structure: space-time.

It was this space-time insight that concerns us here. Einstein's obvious genius won him the freedom to dig much deeper into theoretical physics. It took ten hard years, during which he had to master a highly abstract, cutting edge branch of mathematics before he solved the problem he had presented to himself. The result, published in 1915, was the General Theory of Relativity.

It was an immediate scientific sensation — not only in war-time Germany, but among the scientific community of its Allied enemies as well. This miracle of advanced mathematics gives us insights into much of what we now know about the universe. It tells us of our beginning in the Big Bang. It predicted Black Holes. It explains, with a new, extraordinary precision, how the planets move around the sun. The General Theory of Relativity is even essential for today's everyday tools, such as GPS navigation and your cell phone.

Einstein had unified space, time, energy and gravity: the rate at which time flows depends also on energy and gravity — it depends on the *bending* of space and time by energy, and its manifestation as gravity.

# THE TREK CONTINUES

Einstein showed that the Earth's gravity is a kind of deep well, and anything trying to climb out it needs energy to reach free, unbent, space. He taught us that radio waves and gamma rays shift in frequency depending on whether they are going up or down. Even over a distance of hundreds of miles in altitude the frequency of these waves varies by only a small amount. Here's where The Mossbauer Effect enters the picture. The Mossbauer device is so sensitive that, with it, even the tiny changes in frequency within the restricted distance of a laboratory can actually be measured.

In 1959 Robert Pound, and his student Glen Rebka, used Mossbauer's newly published idea to test General Relativity. The frequency shift they measured in the laboratory matched the prediction of General Relativity.

It is a tribute to Mossbauer's genius that such a simple, but elegant, experiment of very high precision could be performed so shortly after the release of his dissertation.

Normally, before journal publication the results of an experiment are closely held. But there are almost always rumors. In the case of a particularly important experimental result the rumors can get pretty wild. In any event, they reached Caltech and that's why Rudy Mossbauer was invited to become a post-doc.

I suspect that Mossbauer was rather shy. If so, it would be expected when a fresh PhD, especially one who had been rejected by his professor, finds himself walking hallowed halls filled with Nobel Laureates. People did not pay Rudy much attention. That is, until one day he happened to bump into Johnny Pellham.

"Where have you been?" Pellham asked. "I have been expecting to see you, but you never come to see me."

# THE MOSSBAUER EFFECT

It was natural for the two of them to get together since Johnny Pellham was head of low temperature physics at Caltech. The accuracy of Mossbauer's Effect depends on making things very cold.

Rudy replied that he had come almost every day to Pellham's office but he couldn't get through the door. The problem was Pellham's secretary. She was carefully rationing Johnny Pellham's valuable time. Only the most important people gained entrance. Clearly she was not at all impressed with youthful Rudy.

I suppose such shielding was necessary, for many people found excuses to visit Pellham. It wasn't so much their esteem for Pellham — though that was great, and well deserved. No, it likely had more to do with his secretary. Pellham was legendary for having the most beautiful secretary in all of science. I later knew the stunningly gorgeous Lilly Fong after Professor Pellham had migrated to U. C. Irvine. She may have been with him at Caltech. Whoever it was there, she must have been a knockout.

One thing always remains true about exceedingly beautiful women: they are intimidating to a young man. Lilly Fong certainly was to me.

Whether that was the problem or not, Pellham solved it in his usual matter-of-fact way. He introduced Rudy to his secretary with instructions that no matter the time or circumstance, no matter who might be in Pellham's office, Rudy was to go right in. This made Rudy a VIP as far as the secretary was concerned. It was a special honor, given the importance of the people who visited Professor Pellham. I am aware, from my own experience, of the potentates that would cross continents and oceans to meet my boss. In fact,

## THE TREK CONTINUES

I was fortunate to be introduced to some of the legends of physics while working for Johnny Pellham. Rudy now had the opportunity to become widely known among those who really counted in physics. This was to serve him well.

But still, at Caltech he was simply one more post-doc — a talented one to be sure — among other young talents. In these days of monster-sized universities it is hard to realize that Caltech is still the same small college it has been for most of its existence. Even with recent growth it has only about 2500 students, 1500 of whom are graduate students. The Physics Department is historically important, though still small, so everybody there almost certainly knows everybody. Probably, Rudy would also have been known to some outside of this department, as well — particularly because of what was appearing in the literature.

About the time Mossbauer arrived at Caltech, the experiment done the year before by Pound and Rebka finally was published. It must have caused a significant stir in the world of physics and Mossbauer's reputation obviously benefited.

But in Physics things most often take time. A year was to pass before anything notable happened. Mossbauer continued with his research and the usual discussions with Pellham and Feynman. All seemed normal. Nobody was paying much attention to Rudy. But something was going on behind the scenes.

One day in 1961 Johnny Pellham and some of his friends were dining at the far end of the Athenaeum — Caltech's faculty club. It helps to picture the place. It is a large, luxuriously paneled room, elegantly appointed in the style of

## THE MOSSBAUER EFFECT

East Coast wealth — redolent of old money. One always spoke in hushed tones in such a place.

That day in walked Dick Feynman. He paused for a moment until he caught sight of distant Professor Pellham.

"Hey, Johnny!" Feynman bellowed, with his inimitable Queens accent. "Did you hear the news?"

It was a profound violation of protocol, of course, but it was pure Feynman.

Pellham responded in kind, yelling back across the long room.

"No, Dick, what's the news?"

Feynman's turn. He was sporting a big grin.

"Rudy Mossbauer just won the Nobel Prize!"

Dead silence. Then a growing whisper grew ever louder. "Who is Rudy Mossbauer?" everyone wanted to know. "Oh, just some post-doc," might have been the response, though I suspect the reaction was a bit more animated than that.

In any event what had happened was almost unheard of in the annals of physics. A post-doc, only three years after his first, and thus far only, publication had just won the Nobel Prize. Usually decades elapse between theory, experimental verification, and The Prize.

Caltech now had a problem: The next morning the press would be hounding them for information about Rudolf Mossbauer and his position at the school.

Consternation best describes the situation. Mossbauer, for all practical purposes, was little more than an advanced graduate student. And still an outsider to boot, since he had only been at Caltech for a year. How could such a novice suddenly gain ascendency over the vast majority of the faculty? There was not just a matter of institutional prestige.

Much more important, this was a matter of *vanity* for many members of the faculty.

Johnny Pellham and Dick Feynman were having a good time with all this. Particularly since neither of them had a high regard for the Chairman of the Physics Department.

Still, the school had a crisis it had to resolve. And it had to resolve it before the next morning. The faculty went into emergency session. It lasted most of the night. The resolution, Pellham told me, was that Rudy Mossbauer, because of his undoubted brilliance, had been invited to Caltech to assume a full professorship upon arrival. Therefore, Caltech was proud to announce that one of its most esteemed faculty members had just won the Nobel Prize. So much for academic integrity.

Yes, I know that there are varying reports of what happened. There is one that says that Rudy had to wait another year to become full Professor. I like Pellham's version better. Given the nature of academic politics and the dissension in the Caltech Physics Department at the time, it makes more sense. Besides, Johnny Pellham was there! In the center of it all. He should know.

In engineering there is an expression: to "pulse the system." When you want to find out how an unknown device behaves, you send it a pulse — a disturbance — and see how it responds. In this case the *Mossbauer* effect certainly pulsed the institution of Caltech. With all the chaos that ensued, Johnny Pellham, Dick Feynman and their closest friends were highly entertained.

## GOLD FEVER

CALL HIM ROGER. I met him at a party. Intelligent, polished and charming, he soon captured my attention and we wound up in a spirited conversation. A couple of beautiful young ladies tried to insert themselves, but it didn't work. They soon drifted off, faces painted with looks of disdain.

Of what Roger said that evening only one thing remains in my memory. He gave me the lowdown about the executive shenanigans that took place in an eminent technology company that had been, until recently, a Wall Street darling. I had been following the saga of that company, and had had some dealings with it, so I was very interested in what he had to say. He knew what had transpired to cause its sudden disgrace because he was a witness — or maybe a participant.

In the previous couple of decades the company had grown like gangbusters, primarily through acquisitions of other large technology companies. In consequence its stock price had been rising rapidly. Then, investor mob frenzy took over and the stock price of the company skyrocketed. Before long the stock's price-to-earnings ratio was a stratospheric eight times higher than conservatively run companies. But the company made a move that puzzled Wall Street. It purchased a mechanical typewriter company.

The odd thing about the purchase of the typewriter company was that every knowledgeable investor did realize that, in a very few years, digital word processors were going to eliminate mechanical typewriters. The cost of digital electronics was plummeting, and primitive personal computers had already appeared, and the obsolescence of typewriters was already underway. So why acquire a typewriter company? I asked the question of Roger since he had been consulting at the executive level in that technology powerhouse.

Roger's answer? It had nothing to do with technology or profits, it was all about image. At the time of this purchase, the typewriter company in question was still quite profitable. To the uninitiated, the purchase appeared to signal continued growth. The corporate challenge was to keep the stock price high long enough so that executives with stock options could profit greatly when their options reached maturity.

Once the executives exercised their options the cat was out of the bag and everyone now knew what had been going on. For the executives in the tech company it no longer mattered that the stock price collapsed when the illusion evaporated. The stock value soon enough settled down to a more reasonable level.

The executives had bilked their stockholders out of hundreds of millions of dollars. Too bad. That's just business. The greedy investors should have known better. This story that Roger told, with some amusement, was my introduction to a totally alien mentality.

Not long after that party I received a call from Roger. He was masterminding a gold mining venture and he wanted me to join his Board of Advisors. Knowing this project had a high potential for success — it was based on a proven and unique new mining method, which had patent protection — I accepted. This was to be my introduction to *Gold Fever*.

Gold mining is a catchall term that covers a wide variety of methods for extracting gold from the earth. In this particular case the method of interest was the Wanzenberg process. Five years before, in 1975, an elderly geochemist, Dr. Fritz Wanzenberg, had received a patent for extracting gold from carbonaceous ore. Roger had enlisted Dr. Wanzenberg and formed an investor group around him. Before I became involved with the project Dr. Wanzenberg had already constructed a working small-scale prototype of a gold extraction refinery.

This testbed refinery looked for all the world like a giant steampunk samovar. Rising from the floor to a height of about eight feet, its exterior was tarnished stainless steel. Like a samovar, various strange fixtures projected out at odd angles and random locations. Most amazing of all, the darn thing actually worked. Dr. Wanzenberg put raw mine tailings into the machine. After a bit of cooking, out dripped drops of molten gold! Wanzenberg was for real.

Wanzenberg recognized that carbonaceous earth can, given the right conditions, burn. Set the carbonaceous dirt mixture on fire in a retort and pretty soon things get hot enough that any gold that is present will vaporize. Fractionate the vapor, cool it to a liquid, and out drips drops of molten gold. Thus, Wanzenberg's method can provide the fuel for melting and then vaporizing any gold that carbonaceous earth might contain.

Other less efficient methods, using dangerous chemicals, had already long been used for leaching the gold from carbonaceous earth. But Wanzenberg's method was safer, cheaper, and more efficient. Unlike these hazardous chemical methods, Wanzenberg's process needed few, if any, relatively harmless additives.

Not just gold can be retrieved this way. Other precious metals can, as well. Clever Fritz Wanzenberg only needed capital to make everyone involved filthy rich. Investors began swarming to the venture like bees to golden honey.

I had very little to do with the project except to show up for meetings. I was one of the resident "experts" that gave credibility to the effort. So I was mostly window dressing. I didn't mind; it gave me the opportunity to meet some very interesting people.

One day a few of us involved with this project gathered for lunch at a fancy Irvine hotel. At the table was a fellow adviser, Dick Freeman. He became a lifelong friend. What attracted me to Dick was his integrity. It was apparent then, and it remained so for the rest of his life.

In addition to Dick the astronaut Gordon Cooper was at our lunch. He was seated across the table from me. Indeed "Gordy" had been the one with the right stuff — he had tamed his wildly tumbling Mercury capsule, a feat nobody thought possible.

Cooper regaled us with stories that validated his reputation as the wild man among the early astronauts. Stories like the time he and one of his buddies raced each other across the Florida Everglades, so low that their propellers trimmed the swamp grass.

Then there was the Crown Prince of Yugoslavia. Of course he wasn't the real crown prince, though Roger called him that. That he was a collateral prince of the Serbian House of Karađorđević I have no doubt. His trained mannerisms proclaimed his royal heritage. If things had been different, he might, in reality, have been in line for the now non-existent throne.

The prince had a peculiar studied grace about him, as if every movement was rehearsed beforehand and then precisely, and smoothly, executed. His head was always posed to show a regal profile, not condescending but never acknowledging familiarity with his inferiors. His walk was more a glide than the typical American jounce. His arms never reached out without deliberate intention.

He didn't just pick something up, his hand would flow to the desired object and his long manicured fingers would smoothly fuse together with the object. This elegant reach and touch was to cause him a problem when we dined.

The prince had a man servant. This aide was more the prince's shepherd than his servant. You see, the prince was not exactly brilliant. In fact, he gave the impression that he was elegant on the outside but rather empty within. His shepherd had to guide his every action, especially since he knew little of English and even less of American customs.

At noon we took the prince to lunch. The restaurant was a popular Newport Beach family eatery, famous for its high quality hamburgers. Naturally, we all ordered hamburgers and fries. After a short conversation between the prince and his shepherd the prince also ordered a hamburger. I believe his thinking was along the lines of "when in Rome..." What he got was obviously not what he expected. He looked at the burger on the plate and was visibly dumbfounded. How was he going to eat this whatever it was. Gracefully he reached for the silverware, then he hesitated and appealed for help to his shepherd.

## THE TREK CONTINUES

His shepherd informed him that the proper etiquette was to simply pick the object up and eat it. The prince must have thought us all savages. Gamely, and with obvious mental struggle since this violated all his training, he gingerly picked the thing up and took a bite. Too small a bite. The juice dribbled down his chin. He was embarrassed as he wiped his chin. Then he looked around the table at the rest of us happily chowing down, shrugged his shoulders and, now being in "Rome," devoured the hamburger like a true American. One has to learn sometime.

THERE IS SOMETHING ABOUT GOLD that drives some people more than a little nuts, and gold fever was about to infect Roger's project. Things began to get crazy.

Since Roger was my main source of information, I inevitably saw most of what was happening through his eyes. Still, I was beginning to pick up bits and pieces on my own, usually after meetings when, in passing, I would occasionally catch not-so-subtle sarcasms. Factions were beginning to form. Some of the advisors and investors now had doubts about Roger. A few even went so far as to question his legitimacy, and even his honesty. These were sophisticated people who would have known the danger in factionalism. What was going on?

Roger was picking up on this as well. My concerns started to crystalize when Roger told me that a group of investors was trying to take over the project and squeeze him out. Gold fever was running rampant. It got worse. Roger was showing increasing paranoia. When he confessed to me he was now sleeping with an AK-47 under his bed, I decided to bail out.

The decision made, I called Dick Freeman to tell him I was through with Roger's project and why. Dick had independently noticed the same frictions, and growing paranoia, and had independently made the same decision. This congruence of mind was what cemented our friendship. We kept in close contact through the remainder of the crisis. However, both of us continued to talk with Roger since we had not told him our decision.

Our wisdom was confirmed a few days later when I learned from Dick that in the dead of night, a group had broken into the laboratory and stolen Wanzenberg's apparatus, apparently with Wanzenberg's assistance. Frustrated Roger was left high and dry. With no recourse left, he drifted off to other endeavors.

The months rolled by. Occasionally I would get a call from Roger, more just a what's up call than anything. After about a year he called to tell me there was an opportunity if I could create a device which could instantly read fingerprints. Such a device could be used as a key to open a high security door. I thought about this for a couple of minutes and said there was a way to do that. But I wanted to do some tests to make sure.

After the call I rummaged in my junk box and brought out a small prism. The idea I had was to use *frustrated total internal reflection* to capture an image of the fingerprint. I was positive this would work but a test was needed to verify the concept. This was easily done.

The principle is simple. Light traveling inside glass will completely reflect off the surface when it bounces at a shallow angle. This is called *total internal reflection*. However, if something touches the surface from the outside the light will leak out. This is *frustrated* internal reflection. I thought that under the right conditions a finger's ridges would act to frustrate a beam of light bouncing around inside a prism. The fingerprint should be highly visible. This indeed turned out to be the case. What I really wanted to know was whether or not there was any residual trace of the fingerprint when the finger was removed. My concern had to do with the skin oils left behind on the glass. A simple test showed that there was no problem and fingerprints could be reliably seen, erased, and re-established under all reasonable conditions.

A few days later Roger called again. I told him that instantaneous fingerprint detection was easy and such a product could be relatively cheap. If the idea had not already been published or patented then it could get patent protection. I had no interest in filing such a patent. By this stage in my career I had had enough experience with patents to know that they are expensive, time consuming, and generally of no value unless you were trying to protect a business venture. In this case I wasn't.

Roger asked me to sketch up my idea and send it to him. I would receive a suitable reward if his venture panned out. This I did, and heard nothing more about it.

A couple of years passed by. Then, in the trade press I saw a fingerprint reader, made by a local company, that obviously used frustrated total internal reflection. Good for them. Smart people.

A few months later Roger showed up in the trade press. He was involved in a nasty fight with the fingerprint company. He claimed that they had infringed on *his* invention and proprietary rights. Apparently he had somehow previously been involved with this company. I had no way of knowing the details but I had long since given up on any further dealings with Roger. So the whole affair was of only passing interest to me.

Time passed. Then Dick Freeman mentioned that Roger was now in prison. For what, neither of us knew. It couldn't have been about the fingerprint controversy. That was a civil dispute.

More time passed and the probable cause of Roger's imprisonment became evident. Intelligent, polished and charming Roger had methodically stripped away every penny from the elegant, polished and innocent prince. The now destitute crown prince committed suicide.

# FREEMAN

**WE FIRST MET DURING** the gold mining venture. Dick Freeman, like me, was an advisor to the project. Like me, he was an aerospace engineer. The two of us stood aside and watched together while the project spun down into the paranoia and chaos of the gold fever I described in the previous story. Though the venture crashed leaving no trace, for me it left something quite valuable: the treasure of a lifelong friendship.

The rapport between Dick and me was instantaneous. In time Dick's family even made me into something of an honorary nephew. There was Dick, his wife Jane, and, of course, their dog Spot. Even though the two of us eventually lived a hundred miles apart, Dick and I stayed in close contact through long phone calls. I made visits to their home when I was in the Newport Beach area. This continued for nearly forty years until Dick passed away.

So, as you can see, Dick became a dear friend. When we got together, aside from simply enjoying each other's company and sharing meals, as friends do, we talked about things — all kinds of things. Mostly just chit-chat to pass the time. We talked about the past, the present and the future. We talked politics. We talked about family and friends. We talked about our work and our hobbies and interests. We talked about Dick's adventures as an aviator and his abiding interest in all aspects of aviation.

Dick was a dozen years older than I. When we met my career was still in its relatively early stages, Dick was at the top of the heap, and extremely well connected throughout the industry and government. He was senior vice president of a major aerospace corporation. More significant was what he had done to reach that height at an exceptionally early age.

As a young engineering manager at Hughes Aircraft, Dick was working on an air-to-air missile when he had an idea. Why not adapt the missile so that it could attack ground targets. Dick's management turned the idea down. Dick didn't give up. Anyone who is going to succeed in a large organization makes friends with the boss's secretary. So Dick had a talk with the secretary of the Boss — Howard Hughes. The secretary took Dick's one sheet proposal to Las Vegas and slipped it through the crack in the slightly opened door of Howard Hughes' hotel suite. A couple of minutes later the sheet reappeared with the Boss's signature. Thus began the Maverick missile program. And thus began Dick's rapid rise to the top of the industry.

With Dick Freeman's foresight, Hughes Aircraft was more than ready when the Air Force solicited proposals for a ground attack missile. In fact they were already far along in the missile's development. The Maverick missile became the greatest money maker in the history of Hughes Aircraft. It is still in production more than half a century later.

With Dick's abiding interest in missiles and missile technology, he was fascinated by one of my unclassified projects. This was the Hover Test Facility at Edwards Air Force Base. Dick was familiar with the aviation side of Edwards, but the rocket side of the base was new to him. At the time I was earning my living as a consultant. One of my clients had called and put me to work on this project. Dick closely followed my work on this effort while it was underway.

At this time, President Reagan's Strategic Defense Initiative, nicknamed the "Star Wars Program," was in full flower. This program required an interceptor rocket test facility at Edwards. The novelty was that the rockets were going to fly *indoors* inside a large hanger. My client had the contract to precision track the flight test vehicles. I thought this idea was nuts. So did Dick. However, the Government was serious.

These rockets are miniature kill vehicles. Their purpose is to collide with the warheads of ballistic missiles when these nuclear weapons are still in outer space. The facility was intended to test the maneuvering ability of the kill vehicles by flying them in controlled hover. After substantial argument I was half convinced that this thing would actually work, and signed up.

I ended up, off and on, spending weeks in the desert far to the east of the aviation test area. The new Hover Test Facility was to be located on top of a ridge where, early in the Cold War, the rocket engines for the first generations of liquid fueled ballistic missiles had been tested.

My client had already designed the cameras and computers for the rocket test hanger. What they needed from me was some way to measure the vehicle's movements, in real time, with millimeter precision, over a volume the size of a basketball court. These measurements were intended to be sent to the test rocket to control its flight. The task was impossible, yet it had to be done. "You do it," they said. So I did. It required the invention of new algorithms and a completely new way to use the cameras. The project was a great success and the facility is still in use. Rockets *really do* fly inside that hanger!

Of course Dick well knew the aircraft test portion of Edwards. And he knew many of the key test pilots who had been there, the famous and not so famous. All of them he respected for their technical ability and courage, for test piloting is a *very* dangerous business. Most of the pilots he knew he liked a lot, but about a few he had caustic comments. I was more than surprised at how salty he could be in his descriptions of a couple of the more famous among them. I later discovered that most of the aviation community agreed with Dick's sentiments.

One thing Dick talked about with great affection was Purdue University, having grown up immersed in the community. His father had been an administrator in the agriculture school for forty-three years and was the Associate Dean of Agriculture for the last thirty years of that service. Verne Freeman was an *institution* at Purdue.

It is when I met Verne that I began to understand the source of my friend's integrity. It ran in the family. Verne was in southern California to investigate the new wind farm at Tehachapi, and of course to visit Dick and his family. Dick gathered a few of his friends for the wind farm expedition. We traveled in a recreational vehicle the size of a bus. The RV was more a luxurious land yacht than anything, so we rode in quiet comfort while we talked away the hours to Tehachapi.

In those days, the nineteen-eighties, the Tehachapi wind farm was just getting established. It was small potatoes compared to today's installation. Small though it was it was already showing the problems that beset today's wind farms. The biggest problems are two: the lack of reliable power generation and the unreliability of the hardware. Damage to the environment later became an issue with the increasing size of the wind turbines.

Reliable power generation depends on reliable wind. Even at windy Tehachapi Ridge the wind was too variable. The wind had to be Goldilocks just right. A wind speed that was too low would not deliver significant energy to the turbines. In this case it was best to shut everything down to save wear and tear and maintenance cost. Or, the wind speed might be too high and this would drive the turbines to destruction. In this case it was essential to feather the props and shut everything down to save destruction of the machinery.

The second major problem was the stress put on the mechanics of the turbines under all conditions. Even under ideal circumstances the massive gears and bearings tend to wear out quickly, and sometimes even catch on fire from friction heat. Indeed, we saw many broken wind turbines standing forlornly on the ridgeline. In those days operation of the wind farm was just not economical enough to make a profit. Government subsidies were required. They still are.

Disappointed as Verne was with the technology of the time, the information was still valuable. He took this useful information back to Purdue.

Many years later I was visiting Dick when he announced that he had just received the inaugural *Outstanding Aerospace Engineer* award from Purdue University. Great news for Dick. It was certainly worth celebrating. This was in 1999. Dick was particularly pleased that he shared this new award not only with Neil Armstrong but with his best friend Bud Mahurin. I believe he was more pleased about Bud than Neil. Through Dick I had met Bud Mahurin, the famous Second World War fighter ace. So I, too, was glad that Bud had also received this prestigious award.

It was only natural to be impressed with this great, and mild-mannered man — Walker "Bud" Mahurin. Mild he may have been, but in doing the job that he is famous for he was extremely lethal. He had killed many men in aerial combat. Bud was one of America's greatest fighter aces. He was our first double ace in Europe and became a quadruple ace there before being transferred to the Pacific. His transfer came after he was shot down over enemy territory and was rescued by the Underground. He was transferred because he knew too much about the Resistance, and they couldn't risk him being shot down and have the information tortured out of him. In the Pacific, and later in the Korean War, Bud also had multiple kills. He had the unique distinction of having been the only person to have been shot down in all three theaters of war.

Bud, Dick and I shared several meals together. It was at those meals that I learned something of what Bud had experienced. There was exhilaration in winning a fight, of course. But combat had its downsides, as well — it could be very scary, and you could be shot down and tortured, as Bud was. Or, you could be killed. Bud told me that on his first mission over enemy territory he was so frightened he soiled his flight suit. He was terribly embarrassed when he got back to base until he discovered that his condition was commonplace among rookie pilots returning from their first serious mission. No one paid any attention to his condition.

Towards the end of his life Bud received an honor that especially pleased Dick: he was chosen to be Honorary Marshal of the 2007 National Memorial Day Parade in Washington D.C. The year was particularly significant because it was the 60th anniversary of the Air Force. Bud had been selected because he was regarded as one of their greatest heroes.

In between consulting contracts, when I wasn't busy trying to generate new business, I continued my research into the mathematics of neural holography and neural networks. One new discovery became an important algorithm. When applied to an aircraft's image the algorithm could almost instantaneously determine the type of aircraft detected, along with its range and orientation. The algorithm had been validated with field test infrared aircraft imagery.

When I showed this work to Dick he immediately recognized its significance, and arranged a dinner meeting between the two of us and an executive vice president of Hughes Aircraft. The meeting was to be at the posh Balboa Bay Club. What resulted was unexpected, but very educational.

The delicious dinner finished, I was rehearsing in my mind the pitch I was about to make. Only dessert remained before I would be on stage. But before I could begin, we were interrupted. The president of a large company that supplied Hughes Aircraft, and the rest of the aerospace industry, wandered over to our table and greeted his friend from Hughes. The Hughes vice president invited the newcomer to join us for dessert. The fellow declined the dessert but he sat down at our table anyway. The two of them started a long conversation, totally ignoring Dick and me.

I was well aware of the discourtesy this represented to Dick — the guest had shoved aside the host and had taken charge. As for me, I was a nobody so why should these high ranking people care about me, even though my work was the reason for the dinner meeting in the first place. It was soon evident that I was not going to get my chance to show what I had discovered. On the other hand, I was learning from these elitist elite. And what I learned was indeed educational.

Where should the supplier's new factory be built? That was the topic of the conversation between the two of them. Demand for the supplier's product was so great that a large new factory was essential. So where should the supplier put it? There were only two options: Southern France or Northern Spain.

I understood that the tax and regulation situation in California precluded building the factory locally, but there were plenty of places in the United States that had even more favorable tax and regulatory conditions than either France or Spain. But that wasn't the point. What was important was kissing up to the newly forming European Union. It was the evident opinion of these corporate mucky-mucks that America, American workers, and America's future were of no consequence compared to this shiny new toy called the European Union.

Dick was scathing in his comments during the drive back to his house. The discourtesy to me was at the top of his list, but the implicit scorn for America by these senior executives had him fuming. What I took away from the incident was a preview of the reasons for the political realignments that were to take place early in the twenty-first century. If this was a representative example of current top level corporate thinking, it was evident that this self-important class had little interest in, or understanding of, what really made America tick. These were Johnny-come-lately business-school types. They weren't business builders like the engineers who had created their companies. These were the wrong people to be running the show.

## THE TREK CONTINUES

For part of his career Dick was responsible for foreign sales. Of course you make your best sales where there is the most trouble so Dick knew his way around the Middle East. When the government of the Shah of Iran fell Dick helped settle many of its senior members, including the Shah's family, in Newport Beach. He continued advising that community to the end of his life.

Dick Freeman devoted much of his free time to various charities and to helping people individually. An important figure in the Newport Beach community, he served as President of the Exchange Club's Newport Beach chapter. The Exchange Club is a charity devoted to the education of disadvantaged children. At the national level of his charity work Dick was President of the Exchange Club's Western District. A third level Eagle Scout — the highest attainable — Freeman was also a member of the National Council of the Boy Scouts of America.

As close as I was to Dick, until his memorial service I never realized how important he was to the community. One by one, out of the very large crowd in attendance, young men and women came forward. Each testified about the help that Dick had given them through personal counseling and other means. Each said that Dick had profoundly changed their life for the better. Richard Freeman was a very good man.

The memorial proceedings ended with the formal, and very moving, Marine Corps flag ceremony. Dick Freeman had been an artillery officer during the Korean War. Once a marine always a marine. The memorial service filled a large chapel and had an overflow crowd outside. I was sitting in the back where I had a good view through a window of both the outside and the large assembly in front of me. Thus I was aware of the preparations for the rifle salute. The audience in the chapel, facing forward, was not aware. At the sudden and unexpected crack of all the rifles, the entire crowd, in perfect unison, reflexively bounced an inch or two into the air. Given Dick's sense of humor, he would have been highly amused.

# LIFE SAVERS

"Bill, you are going to die. You will be dead within the year unless you stop this nonsense. Change and you will have at least a chance to survive." Bill O'Neil, the mentor I spoke about in the story Failures, was smart enough to take the warning seriously and make appropriate adjustments. He was determined to stay alive. I was relieved when he told me this for Bill had become one of my greatest friends.

Bill's problem was his addiction to tobacco. Each morning he brought in a fresh carton of cigarettes and stored them in his desk. He also kept a couple of unopened cartons in the bottom drawer, just in case. When one cigarette burned down he would use it to light the next. This continued through the day, with one exception — Bill was the soul of courtesy. When a non-smoker, such as myself, entered his office the cigarette was immediately extinguished. The strong air conditioning quickly cleared the air and made the room habitable for the visitor. He lit up again when the encounter ended.

Everyone tolerated this addiction for two very good reasons: Bill was a genius. And, he was the most congenial of men. We all delighted in the play of wit when he was around to stimulate things.

You may remember I had met this remarkable man at the Aeronutronic plant of Ford Aerospace. One of the reasons

## LIFE SAVERS

I admired Bill was he was wide open. The research I had been doing was completely new to him. His background was aeronautical engineering, not my world of neural networks as they are now known. What I was telling him had to have been very strange. But Bill instantly picked up the essence and started creating on his own. It was the beginning of his conversion from the analog world of aircraft flight controls to the digital world of image processing, where his influence was enormous, and ongoing.

During the first couple of years after I started working for him, Bill gave me a variety of assignments which moved me around the company. For a while I was the Chief Engineer for Strategic Systems. Then Bill created the Digital Design Department and brought me in. My first task was to identify and recruit the Division's top technical talent. The challenge was substantial.

I solved it by telling people the truth: "If you go to work for Bill you will lose money. The annual raise pool is the same for every department. Where you are now you get your department's highest raise. With Bill you will be competing with the best in the company, so your raise will be small. The good news is you will be working *with* the best in the company, and you will be working *for* Bill O'Neil. It's going to be very entertaining. And, you will learn a lot."

My approach worked every single time!

When I first met him Bill already had made two major contributions to modern flight technology. A fresh MIT graduate, Bill was running the wind tunnel at MIT when the great aviation pioneer Ed Heinemann scooped him up and put him to work. It was there Bill accomplished the lesser of his first two major achievements — the autopilot for the

A-4 Skyhawk fighter jet: "Heinemann's Hot Rod," the favorite mount of the Blue Angels, and the most maneuverable aircraft of that generation. Bill's creation was a highly influential landmark in aircraft flight controls.

Bill's second achievement changed world history. Aeronutronic was the nation's leader in high speed reentry vehicles — the devices which carried ballistic missile warheads to their targets. During the *Reentry Measurements Program* Bill had the idea that multiple reentry vehicles (RV's as they are called), launched from a single rocket, could be accurately directed to independent widely spaced targets. The program tested Bill's idea, bus design, and equations, and the rest was history. This was the genesis of the Multiple Independently-Targeted Reentry Vehicle (MIRV) ballistic missile. MIRV ICBMs completely changed the strategic balance between the US and the USSR, and was a major factor in the fall of the Soviet Empire.

There was much more to Bill than just his superb technical creativity. He was a wonderfully subtle, and practical politician. For years he had made budget requests to establish a digital image lab. The requests went nowhere. He solved the problem by sending me to assist the Business Development Director. I did the job with impeccable attention to the director's desires — with one exception. I inserted a budget item for the digital image lab. "Why this?" He wanted to know. I replied that digital brains have to have eyes. The lab provides the eyes. The lab was approved.

Bill had other ways of solving problems. In those days Aeronutronic produced the Sidewinder air-to-air missile. This weapon, created by the Navy in the late nineteen-forties, was still, in the nineteen-seventies, rather primitive. It had proved itself in combat but it was expensive to produce and not very reliable. Bill wanted to know why it was so expensive and unreliable. So he paid a visit to the missile people. What he discovered was that much of the cost was in the missile's calibration procedure. The analog circuitry had to be fine-tuned with many precise adjustments being made just-so for the thing to work. What was worse, the missile still had some vacuum tubes, which lacked stability.

Bill offered to fix the problem but was politely shown the door. "Not Invented Here." He could not get budget from the missile people for the Digital Design Department to work the problem. Okay, Bill simply did the job at home. He had everything he needed there because he ran a side business supplying digital controls for the petroleum industry.

Some weeks after his rejection he showed up at the missiles shop with a digital card. He sweet talked the folks into allowing him to plug the card into the calibration test rig. With a single turn of one potentiometer Bill brought the missile into full calibration. A five minute test job replaced weeks of expensive calibration. What's more the new thing had no vacuum tubes. It was completely stable. Skeptical missile people became true believers, and the Sidewinder missile went on to great success.

FUNDAMENTAL RULE: NOTHING LASTS. At the end of the 1970s Bill left to become the Chief Scientist at Autonetics.

I was asked to interview his potential replacement, who charmed me with his personality and technical knowledge. Boy, was I fooled. We all were. This guy turned out to be a totally insufferable con man and bully. He understood that to be a member of the club you had to know the jargon. So he had memorized the lingo, but was absolutely clueless as to what it meant. Within days the department started falling apart. A few months later this arrogant individual stepped on a senior executive. The executive called a guard who immediately escorted the miscreant out the gate. But by then I had already left the company.

As so often happens, being confronted by this character turned out to be a blessing in disguise: his presence propelled me in a different, and, it turned out, better direction. Before he was booted out, I realized there was no way I could continue working for him. Besides, the handwriting was on the wall. Ford had already shut down the research department and was beginning to sell off much of its technology and manufacturing to Hughes Aircraft. Shortly thereafter Ford closed down the division and sold the facility. The property on Jamboree Hill had become more valuable than the company. Not much more than a dozen years before Aeronutronic had been one of America's technology jewels. It exceeded TRW in capability, reputation and growth. Then Ford bought the company, looted it, and tossed it into the trash.

I could have followed Bill to Autonetics but the commute was daunting. Instead, I decided to try my hand at consulting. A contract with Autonetics got me started. By dint of very hard work, and more than a few miracles, I survived in this most difficult way to earn a living.

Bill lasted a couple of years at Autonetics, then teamed up with the Director of the Army's Night Vision Laboratory to found a new company. The enterprise was a great success and was soon snapped up by Westinghouse. Bill supplied the technical brains and the former director ran the business. Its product was the most amazing infrared camera. This was in the early 1980s when such cameras typically produced fuzzy images. Bill's creation produced broadcast studio quality imagery that wasn't equaled elsewhere in the industry until well after the turn of the century. Sadly, his camera was so good it quickly vanished into the Black World.

Bill had hired me to write the specification for the patent application. For years he had been my mentor and my friend, but I had only seen Bill from the outside. In writing this specification I had to master the technical details of his invention. Now I saw him from the inside. I saw how his mind worked. In the technical world the highest praise is reserved for creations that engineers describe as "elegant." This is a solution of such simplicity, effectiveness and beauty that it is instantly recognized as the ideal — a true work of art. Bill's invention was elegant.

The sale of Bill's company had given him financial freedom. He migrated north and settled in the forests of Oregon. When Northrop-Grumman bought Westinghouse Bill became a key figure at the highest levels in that organization. It was at Northrop-Grumman that Bill made his last great technical contribution.

Lockheed and Northrop teamed up and won the F-35 fighter competition. Among the fighter's many virtues was something revolutionary: the Distributed Aperture System, or DAS, as it is commonly known. Spaced around the F-35 are six high resolution infrared cameras. Digital processing seamlessly merges their images to create a comprehensive spherical view of the world. Night or day doesn't matter, with infrared sensing the surrounding world is always visible. This imagery is delivered to the pilot through his helmet.

Wherever the pilot looks he sees this imagery as if he had magic eyes. The walls of the aircraft vanish and he seems to be floating in space, able to look down through the floor or at what is behind the plane. In short, the pilot now sees what is going on over the entire sphere surrounding the aircraft. Moreover, the merged imagery is also used by the aircraft for automated battle management, both for offense and defense. Using the latest missiles the pilot can even attack enemies that are chasing him.

DAS has revolutionized air combat. In that arena situational awareness is vital for victory. DAS gives F-35 pilots supremacy in situational awareness. As a consequence, in mock combats against the best of the best, flying older aircraft, the F-35 wins every time. DAS is now considered essential for many of the next generation of military aircraft. Once exposed to what DAS offers pilots no longer want to fly their old aircraft. For a fighter pilot DAS is a real life saver.

DAS, the invention of William F. O'Neil, is a technical tour-de-force.

In the years after Bill moved north we remained in close contact by phone. Long conversations took place almost on a weekly basis. Of course he couldn't tell me much about his work, that was both proprietary and classified, but we could still talk technical and he was always interested in what I was doing. Besides technical there was a whole world of fascinating ideas to explore.

A few years ago, after a long conversation about the ins and outs of the DAS, and another project about swarms of robot aircraft, Bill mentioned that he was going into surgery to correct a hernia. I gave him a couple of weeks to recover then phoned him to see how he was doing. His voice on the phone

## LIFE SAVERS

was strong. But it wasn't him. It was his son, who sounded just like him. Bill had died the day before and his son was making funeral arrangements. The surgeon operating on his hernia discovered that Bill had fast-acting pancreatic cancer.

ONE DAY LONG AGO I walked into Bill's office at Aeronutronic just as he was removing the cartons of cigarettes from the bottom drawer of his desk and tossing them into the wastebasket.

Into the drawer went cartons of Life Savers®. They'd given Bill another 40 years. Thank God.

*Space shuttle Atlantis is seen as it launches from pad 39A on Friday, July 8, 2011, at NASA's Kennedy Space Center in Cape Canaveral, Fla. The launch of Atlantis, STS-135, is the final flight of the shuttle program, a 12-day mission to the International Space Station.*

## THE ROCKET SCIENTIST

FOR DECADES SOME PEOPLE have called me a rocket scientist. Really though, I have never thought of myself as such. Whenever I'd hear that description I would immediately change the subject.

True, much of my career involved rockets of various kinds and the satellites they lofted into orbit. Once I even designed an instrument package that flew into space. However, I had good reasons to not respond to the label. I knew and worked alongside the real thing. Knowing and admiring real rocket scientists, I was not about to promote myself as one of them. I was not really a member of their club — people who designed and built rockets for a living.

Now a rocket has many parts. For example, a liquid-fueled rocket has a payload section, a structure, and propelling engines. It also has guidance and control mechanisms. And such a rocket is a system that requires integration of those elements. Anyone who works full-time on one of these aspects could be considered a rocket scientist. Come to think of it, since I occasionally worked on guidance and control, I temporarily was in the club. But then, after a brief period, I went on to other things. Consequently, I never considered myself a full-fledged member.

Periodically I did become deeply involved with real rocket science — more as a bystander than anything. For example, there was the memorable occasion when I was in the blockhouse of a rocket test stand when the engine blew up. It caused quite a mess. And was so toxic we had to remain in the blockhouse until it was safe to emerge.

But while the majority of my career was spent pushing equations to understand other aspects of advanced technology — part of it was space satellite or rocket-related. Let me tell you a story about a time when I was more involved, and the true rocket scientists with whom I was privileged to work.

During the 1980s, while working as a free-lance consultant, I had gained enough favorable attention to receive a very attractive offer to join a beltway bandit. These companies provide System Engineering and Technical Assistance (SETA) engineering guidance to the Government. They have major offices in the beltway surrounding Washington D.C.

This SETA wanted me because of my experience with High Energy Laser weapon systems. These "death rays" were a key technology of President Reagan's Strategic Defense Initiative. The program was derided by his critics as his "Star Wars." The name stuck. This particular SETA managed the majority of the Star Wars effort. Here was a chance for my ideas to influence Reagan's initiative.

The offer was particularly attractive because the Los Angeles division of this SETA had hired, from nearby Rocketdyne, the cream of the crop of its rocket engine designers. These were the people who had master-minded the giant, one-and-a-half-million pound thrust F1 engines that propelled the Apollo Saturn rocket toward the moon. They also had created the Space Shuttle's Main Engine (SSME). The chance to work not only with real rocket scientists, but with the true royalty of that profession, was irresistible. Sarah and I moved north and began a new life in a new town.

These rocket engineers were an interesting bunch. By comparison with most of the people in the aerospace industry they were a bit eccentric. I suppose they had to be to take the risks inherent in rocketry. After all, rockets, while under development, have a disturbing propensity to explode. Expensive failure is the norm, and one has to be out of the ordinary to accept catastrophe as the day-to-day cost of doing business.

Seasoned pros on the team included two German experts in turbo pumps. I never asked them, but from their advanced age and deep experience, it was evident they had been part of Von Braun's crew at Peenemunde developing V2 rockets during the Second World War.

As Principal Advisor to the Air Force for rocket propulsion Tom Coultas was the king of the realm of rocket engine experts. At Rocketdyne he had balanced the design of the SSME. This means that the Space Shuttle's main engine could be considered his intellectual creation. Earlier, as head of the thermal design section, he had been a key figure in the Saturn's F1 engine development and its J2 upper stage engine. He had also fixed some serious problems with the Lunar Lander's engine. But there is more than that — a very good reason Tom's colleagues universally looked up to him as *their* Grand Master.

I figured out what he did to have gained such esteem among his colleagues only recently. There were earlier clues, things my colleagues had told me, but some of the puzzle pieces were missing. It took a study of the structure of the Saturn Moon Rocket to realize what a feat Tom had accomplished.

The Saturn V Rocket had a problem with pogo oscillation. I knew Tom had solved a pogo problem, and thought he had eliminated some kind of issue with the thrust chambers of the F1 rocket engines.

I now realize Tom tackled a much more critical problem, and it wasn't in the engine's thrust chamber after all. The problem was the oscillation of the Saturn Rocket's structure. The Saturn Rocket could stretch and contract like a vertical spring, its top end oscillating up and down. This is the pogo effect, something like the bounce of a child's pogo stick. The unmanned Apollo 6 flight test of the full-weight Saturn rocket showed that the pogo oscillation was violent enough to probably kill the astronauts. The Apollo program, which had intended to fly astronauts on Apollo 7, now faced unrecoverable total failure. The national humiliation would have been unbearable.

Then.... *Along Came Tom!*

Realizing the structural pogo oscillation affected the oxygen flow to the engines and this synchronously modulated their thrust, Tom determined the resulting amplification by positive feedback was the cause of the problem. His fix was to put tuned Helmholtz resonators in the oxygen supply lines. This damped out the engine and pogo oscillations so that the astronauts could survive the launch. The program could go forward.

In short: Tom Coultas saved the Apollo Program and America's honor.

But one wouldn't have known that Tom Coultas was an historically important figure. Tom was just Tom. He was unprepossessing. Genial says it better. For example, many were the hours I sat next to his desk talking about this and that. Every once in a while his Irish Setter, lying nearby, would look up at me, yawn, turn in a circle, then go back to sleep. Whenever I had a technical question Tom would spend whatever time was necessary to make sure I got things right. On some occasions he would simply take on the question and do his own calculations on my behalf.

I suppose you could say Tom Coultas was phlegmatic. He had to be, having survived many years of hazard in his profession. But once in a while he would get riled up. A consultant to the State of California, every so often Tom would travel to Sacramento to attend the meeting of the state's Air Resources Board. For a couple of days thereafter he would be in a fit at the idiocies he witnessed at that circus. Eventually, though, he would regain his composure. I understood some of his frustration for I could see through his eyes a bit of the insanity.

That excursion being the exception, it was really hard to upset Tom. He tended to take things in stride even when his colleagues were in a panic. One morning I walked into Tom's office and reported that Ron Cook, in the office next door, was in a frazzle. Tom was probably not surprised. I expect that he heard Ron's more than mild expletives through the partition separating them. Ron was upset because he had just gotten the report that, during its recent Shuttle flight, the walls of the thrust chambers of the Space Shuttle Main Engines had cracked open and hydrogen fuel was leaking directly through the side walls into the chambers.

Ron had good reason to be upset because he had led the design and development of the thrust chamber. Now

## THE TREK CONTINUES

NASA, not having fully digested the management lessons of the Challenger disaster a couple of years before, insisted on running the engine well beyond its rated power level. This way they could put a heavier payload into orbit. Ron's team, constrained by NASA's original specifications, had done an excellent job. But those constraints allowed only a very small safety margin. Now NASA clearly had exceeded its own design margins and Ron's creation was suffering.

Tom wasn't the least bit fazed at my news. "Good," he said, "the thing's obviously working fine. They should leave it alone." He had a point. Tom was a pragmatist. Even when he didn't know why something was doing its job he was perfectly happy to let it work. Of course Tom was an old school engineer from a time when things were developed cut-and-try, and designs were calculated with slide rules rather than computers. (He still preferred to use his slide rule.) In this case, though, Tom had figured out what was probably going on. The engine was redesigning itself to incorporate transpiration cooling. And, as a result, it succeeded nicely in lofting payloads into orbit. But without change, still there would be a risk — the engine was uncontrolled, and they couldn't predict what it would do in the future. Redesign was in the cards.

Time for some inside baseball: When most people think of rocket engines, they are usually visualizing the expansion bell which extends below the thrust chamber because that is the most visible part. Much more important is the thrust chamber, for it is the heart of the engine. Wrapped around the thrust chamber of the SSME is an incomprehensible tangle of machinery that few try to unravel.

The thrust chamber is where the real action takes place. It burns the fuel and oxidizer to create a super-hot, very high pressure gas. Most of the engine's thrust is created when this gas is accelerated through the narrow nozzle of the thrust chamber. The expansion bell downstream matches the gas flow to the outside world and adds a bit more to the thrust.

*Close-up side view of Space Shuttle Main Engine (SSME) 2059 mounted in a SSME Engine Handler near the Drying Area in the High Bay section of the SSME Processing Facility Space Transportation System, Space Shuttle Main Engine, Lyndon B. Johnson Space Center, 2101 NASA Parkway, Houston, Harris County, TX.*

*The problem that engineers face is that the thrust chamber's gas is very, very, very hot. In high performance engines there is no material that can withstand both the heat and the pressure. And with all that heat, liquid hydrogen fuel very quickly turns into gas. The solution is to actively cool the wall of the thrust chamber, and often the expansion bell as well. This is done by circulating the cold fuel through narrow channels built into the wall.*

What Ron was frosted about was the new cracks in the side wall of the thrust chamber allowed a portion of the vaporized hydrogen to leak directly into the thrust chamber from the side rather than being injected properly from the top where combustion could be controlled. Tom, on the other hand, was well aware of a scheme for cooling the inside wall of the thrust chamber by deliberately flowing some of fuel through small openings, or pores, in the side wall. This sets up a thin film barrier of cold gas which prevents the very hot gas from ever reaching the wall. This approach is called transpiration cooling. That method was rejected because it was still unproven.

In this case, Ron's thrust chamber took matters into its own hands and opened up just enough cracks, in just the right places, to establish this transpiration protective layer of cold gas. After the crack formation the engine actually worked better than the original design. NASA, no doubt embarrassed, changed the specifications and had a new thrust chamber designed to handle heavier payloads without the engine needing to fix itself. It was to be one of a great many reworks over the life of the SSME.

Obviously, during my sojourn at this SETA I was learning a great deal about rocket engines. And, overall I was having a good time. It didn't last — the division manager and I had a final falling out. After four years at the SETA I was back on the street and working as a free-lance consultant. It was only then that I began to really dig into the detailed technical aspects of jet and rocket engines and compressible air flow. Tom and textbooks were my teacher. What better way to learn than one-on-one instruction from the Industry's Grand Master of rocket engine design!

Getting booted out of the SETA proved very much to my benefit. I now had the freedom to invent as I pleased. This led to new inventions and new patents when I wasn't busy consulting on projects that fascinated me. After a few years Raytheon presented me with an offer I couldn't refuse: a position as an Engineering Fellow at Raytheon Corporation at, by comparison to what I made at the SETA, a princely salary. I stayed at Raytheon until my formal retirement more than a dozen years later.

*Mom and Dad, Family photo collection*

# FAMILY

I LIKE TO PRETEND THAT I CHOSE my wise parents. The truth is they chose me. They chose me by carefully shaping the raw material of their boy into the man I am today. What they saw as the good they encouraged; the bad they disciplined. And they allowed me the freedom to grow to become myself. With suitable guidance I had the essential liberty to find my own way, to make my own mistakes and learn from them. Year by year as I grew they showed me new ways of looking at the world, and gave me new responsibilities in support of the family.

Mom and Dad's greatest gift was the example of their love and care for each other, and for my sister and me, during the early difficult times and later in prosperity.

*Grandmother Richards (Alta Sayles Richards)*

# HERITAGE

"I WAS BORN IN A LOG CABIN high atop a wilderness mountain crag. My pop was a grizzly bear, my mom a wildcat. My grub was tree bark and rattlesnakes." So begins the prototypical American autobiography. Well, almost.

I won't claim such a beginning. Mine was actually right smack in the center of mainstream America – pioneering and immigrant America. As I wrote in Book One of these stories, *From The Potato to Star Trek and Beyond*, most of my information of Dad's forebears extends back to New England. Mom's parents I know were immigrants from Sicily. As a child I was blond-haired and blue-eyed. That later darkened to Mediterranean hair and eye tones. From the occasional occurrences of this coloration among my mother's relatives we can surmise, but not know for sure, that our family's Sicilian heritage also extends back to the Norman and Viking conquerors of Sicily.

Genetic recombination being what it is, little of those hardy folk still lives on in me. Rather, the Norman Rockwell painting, *A Family Tree*, with ruffians, pirates, scalawags, Indians, dance hall dolls and puritanical preachers – as well as other typical Americans – provides a far more accurate picture of the well-scrubbed American boy that was me. Thus, though only a few stories and fragments of my ancestral

## THE TREK CONTINUES

family history survive, it's enough to show I am a member in good standing of America's colorful Mongrel Horde.

In that previous memoir I told of my mother and father, how they met and how they shared sixty- six years of life together. I told of my great uncle Chet Sayles, after whom I am named, and who looked after the very young me. I told of his father, my great grandfather, Oscar Sayles — a scout for a covered wagon train traversing the Oregon Trail. I told of his later exploit carving through the forest wilderness the road that later became part of the great west coast I-5 highway. I told about my mother's parents and how they crossed the ocean and became proud Americans. And I told of my mother's sisters and other accomplished men and women who preceded me.

Here I'd like to share a bit more about the people in the family whom I got to know in life. and who loom large in my heart and experiences growing up.

My mother had several sisters. Two of them married men who rose to prominence. Aunt Frances married Nicholas. Uncle Nick became prominent on Wall Street by chairing the working group which developed NASDAQ. It was later, in high formal ceremony, that Uncle Nick was dubbed a Papal Knight.

Aunt Josephine corresponded for a long time with Santo, a young Sicilian. She went on a visit to Palermo and came back his bride. Uncle Santo had his own history. His father was in the "import-export" business. I put this in quotes because, except for exporting olive oil, it isn't clear exactly what he imported and exported. It has been reported, though, that there were always a number of tough looking characters hanging around the palace in Palermo. Beyond that your guess is as good as mine.

# HERITAGE

Now, Santo's father observed that his son had a fine mind and good moral character. For these reasons he decided that Santo was not suited to join the family business, whatever that was. Instead, Uncle Santo's destiny was to receive the finest education. He took to study with grace and enthusiasm and turned his talents to medicine. Upon arrival in America he recertified as a doctor and set up his practice among the Italians and Jews of Brooklyn. He thrived, for he was highly skilled and much beloved. Part of the reason for the community's respect for Uncle Santo is that he had this strange notion that sick people should stay in bed, especially sick children. So, into the 1980s he continued to visit the homes of the sick. Better he should be there than the sick should be stuck waiting in a crowded and disease-ridden office, where, vulnerable themselves, they could also spread whatever disease they were fighting to others.

There came a day, finally, when Uncle Santo decided to retire. A couple of years into this new existence he was pacing the house like the caged lion he had become. So, he announced he was reopening his practice. The neighborhood was delighted. The day before his return from retirement, family and his many friends gathered to cheer Santo back into the world. In the evening, full of satisfaction and love, Uncle Santo went to bed as usual. That night he died in his sleep.

It is my opinion that my mother, like her sisters, married exceptionally well. But maybe I'm just prejudiced, for as a child and youth, I was enveloped in love and respect, together with good discipline. I certainly have had a fair share of successes, as a result. My sister, Marie, has done better, with degrees from Harvard, Georgetown, and Berkeley, and a Postdoc at Oxford. As I've noted in *From The Potato to Star Trek*, she served as a diplomat with the U.S. State Department's Foreign Service. Having been stationed in Pakistan, India, China and Afghanistan, she also served on the National Security Council, under two presidents, as Director for Afghanistan, before retiring.

In short, Mom's family settled nicely into the American Dream.

I have written about one of my great-grandfathers, Oscar Sayles. My other great-grandfather, on my Dad's side, was also rather colorful. My father remembered him as a formidable looking old man, nodding placidly in his rocking chair, warmed by the flames in a great stone fireplace, his sword by his side, just in case. Great-grandfather Truman S. Richards was from pioneering Wisconsin, and before that, his folks were from New England. "Too crowded here," Truman's parents decided when new neighbors moved in a mile down the trail. So they picked up and homesteaded a large farm in Wisconsin.

Great-grandfather Richards earned his manhood in the Civil War. Enlisting as a sergeant in the 33rd Wisconsin Infantry, he rose through the ranks to become a first lieutenant. He later settled on a large spread by the North Platte River in Nebraska. My great-grandfather's near neighbor in Nebraska was Buffalo Bill Cody, who lived on his ranch, "Scout's Rest."

Ora Richards was a son from Truman's first marriage. He did very well, becoming, at an early age, the Sheriff of Grant County, Nebraska, a true Western Sheriff of still wild country. But if Ora became Sheriff at an early age, he also died at an early age, following his mother into the ground. Truman was evidently devastated at this double loss. He abandoned his spread and moved to Seattle, with his second wife, where my great-grandmother could be close to her dear sister.

*Great Grandfather Truman S. Richards*

I know little of the son of Truman S. Richards, my grandfather Truman L. Richards. I never met him. He died as a result of an automobile accident in 1929. Dad was in the vehicle but was thrown free. He did not come out of it unscathed; a nearby car ran over his leg. He carried that ugly scar to the grave.

Mom and Dad have now both passed, Dad at the age of ninety-three and Mom at ninety-five. Both declined with grace.

In his last years Dad mostly sat quietly in their comfortable Virginia home, dreaming of long-ago adventures: printer's devil, college student, ranch hand, hobo, roustabout, aerial photographer, army master sergeant, union official, businessman, mediator, printer of currency for many foreign nations, successful inventor and engineer — occasionally telling stories about these adventures and achievements.

Mom's career had a different trajectory. She was a public health nurse. Her nursing years started with a challenging residency at frightening Bellevue, dealing with the insanity there. Then came the war years and marriage. The war took the family south where she nursed people living in the disease-ridden backwoods and swamps of Georgia. The tough and dangerous Hunter's Point District in San Francisco, where she and Dad moved, was little challenge after the hazards of Georgia. Mom continued with public health nursing when we moved to Southern California until she retired to give birth to my sister. After which, she returned to school and picked up a long desired bachelor's degree. What can I say about this good woman? She was as close to the ideal mother as is humanly possible. She is with me still.

## SOME ASSEMBLY REQUIRED

IT USUALLY COMES in a flat brown cardboard box. Your first impression, *How can that bulky item I thought I had purchased fit inside that small package?* Maybe in the advertisement for the product you saw a warning: *Some Assembly Required,* Or maybe, just to falsely entice you, that warning was left off. No matter, you are now stuck with an obvious job of assembly. But it should be easy. You've done it before.

With some struggle you slice and tear at the box until small plastic bags full of little parts start to dribble out. Finally you get the box sufficiently demolished that you can see how some packaging genius had managed to squirrel away something ultimately big into something surprisingly small.

Nowadays the instruction sheet is nothing but a cartoon strip. Seldom do you see a written word at all, and when you do it will be one or two sentences repeated in a multitude of languages and scripts. Usually these words will be little more than warnings about the hazards of plastic bags.

THE CARTOON INSTRUCTION SHEET makes little sense at first glance. Fortunately someone had the good sense to number the steps so at least there is some flow in the instruction. Mostly what you see has lines that direct screws into holes, perhaps with a nut on the receiving end. If you're lucky, eventually you figure out what these cryptic figures are trying to tell you and you start to work.

Of course nothing goes together the way the instructions say — or the way you think the instructions say. You're left to figure out for yourself how the darn thing must be assembled. Actually you get the thing half way put together when you realize that the instruction sheet had the right order of assembly, after all. So it's backtrack a ways and start again — the correct way. After hours of struggle you finally get the thing together and working. You sit back with a sense of triumph at having overcome the impossible.

Strewn around you is the debris of the packaging. Your bones aching from the just finished labor, you summon the effort to gather up this mess and consign it to the trash bin. At last, completely finished, you collapse. You are a little peeved at how much time and effort it took to get this product to a usable state. If only you could go to a store and buy the thing already assembled and working, like the old days. That would save so much time and grief. You would be willing to pay the premium just to save the hassle of *Some Assembly Required*. Alas! Those times are mostly gone. It's mail order or big box stores these days. The small flat boxes marked *Some Assembly Required* are the same from either.

Despite all the grief these days, we should doubtless count our blessings. It could be worse — much worse. My recent experience with *Some Assembly Required* brought back memories of a particular Christmas Eve more than half a century ago. On that day I was a graduate student visiting home for the holidays. Christmas was a time of great excitement for my kid sister Marie — she was only seven years old. A fire blazed merrily in the fireplace the night before the

Big Day. We had erected our Christmas tree and festooned it with lights and delightful decorations earlier in the day, Marie happily dancing around while gathering up ornaments and handing them to the rest of us. Out from the closet came gaily wrapped presents to be strewn around the foot of the tree. Then we sat down for our Christmas Eve supper.

Soon it was time for my drowsy sister to head for bed. Before that, though, we had a family tradition to observe. My sister was allowed, as I had been many years before, to open one gift of her choice. This done, off to happy dreams she went.

After Marie went to bed the evening continued with pleasant family conversation. Then Dad remembered that there was one more thing that had to be done. He had forgotten Marie's doll house. We still had to assemble that. The doll house was to be a special Christmas morning surprise for my sister, which is why it was still hidden away.

Mom left us to sort things out and wandered off to bed. Early to bed, fresh for Christmas day.

Out from the closet came the unassembled doll house, packaged in a flat cardboard box. Inside the box were several colorfully printed thin sheets of enameled steel and the instruction sheet, *Some Assembly Required* at the top, very little more.

That should not be a problem. Dad was a mechanical genius as I well knew from experience. He could take apart the most complex machine, clean the scattered parts, lubricate them, then reassemble the machine so that it worked even better than factory fresh. I had helped him on a couple of the most difficult of these projects so I understood his remarkable gifts for visualization of spatial relationships. Assembling a doll house should present little challenge.

## THE TREK CONTINUES

Dad scanned the instructions, then, expressionless, he handed them to me. What is this stuff, I wondered? I had a degree in physics and two years of professional engineering under my belt but I couldn't make heads or tails out of this compilation of written gobbledygook. There was a single line drawing of the assembled doll house, with a few notations pointing out various items, but that was little help. The text was an almost indecipherable tangle: *"Put tab G-Z into slot 7-3 while being careful not to fold along the line C4 – C4' until subassembly A5 has been attached."*

We were stuck with figuring all this out without much help. We laid the printed metal sheets out on the rug and rearranged them a couple of times until they made some kind of sense. Then we examined each carefully, making sure we knew the labeling for each slot and each tab. This done we worked out a probable sequence of assembly with some guidance from the instructions. Complicating the problem was the need to fold parts of the sheets to create box girders so as to make the doll house rigid. All this had to be done in just the right sequence or the thing would be a botch job.

Long into the night we worked on this most challenging puzzle. Eventually, we triumphed. It was together, rock solid and looking splendid — a miniature antebellum mansion complete with a portico on one side but wide open on the other so that a collection of tiny furniture (still resting under the tree inside gaily wrapped boxes) could be inserted.

The now completed doll house was moved to the place of honor under the Christmas tree. It was time for us weary two to go to bed. We knew it was late at night but we still did realize how late until we looked at the clock. The sun would be rising in a couple of hours. We would get little rest before the Christmas morning festivities.

Christmas Day began with church. Then home it was and time for family celebrations.

After breakfast we gathered around the Christmas tree and began the gift giving. My young sister showed extraordinary patience waiting for the moment when the pretty boxes were to be unwrapped. Of course, she had seen the doll house when she woke up in the morning, but there was also a pile of boxes with her name on them. Once finished with the unwrapping, Marie's delight with her haul of goodies, capped by the doll house, radiated joy and filled our home with warmth and good cheer.

The family cat Sam — orange tabby Samantha — spent the morning pouncing on, and diving under, the scattered wrappings and ribbons before collapsing in complete exhaustion.

A sumptuous Christmas feast ended a joyous day, a day well remembered by all. Wish I had a photo of that day!

*Christmas Day 1942. Dad, me at a little over 1 year old, and Mom*

*My sister Joan Marie, then a young woman of 20*

*Mom in her kitchen*

## MOM'S HOME COOKING

IF GOD WERE TO GIVE ME the privilege of reliving just one moment of my life what would it be? There are so many times that are special that it would be hard to choose. The day of my wedding to my dear Sarah is at the top of the list. When the letter came accepting me to the UC Berkeley that was certainly a moment to remember. The day I was hired onto my first full-time job was quite a thrill, particularly since it was so unexpected. While still a novice kayaker, running Blossom Bar on the Rogue River was an experience never to be forgotten.

Actually, the choice is easy. Of all the treasures in my experience there is just one that stands out above all the rest: I would really love to have one more taste of Mom's home cooking.

But which dish? There are several treats that tug at me, but two rise to the top: Mom's chuckwagon casserole, and her spaghetti and meatballs. With these two Mom reached the pinnacle of culinary perfection. I am left with a dilemna, how could I possibly choose? Please, God, may I have both? I have to choose just one you say? Tough choice, but make it spaghetti and meatballs.

Before I describe these delights let me give an example of the magnitude of Mom's talent in the kitchen. When necessary she could turn something genetically unpalatable into a meal that was at least acceptable.

# THE TREK CONTINUES

When I was very young we were really pretty poor. It often was necessary to somehow make do. Mom, being a nurse, was concerned with how to maintain my health. At an early age I had a number of serious illnesses beginning with a near fatal encounter with bronchial pneumonia when I was an infant. So Mom tailored our meals not only around her small budget but also around my health. I needed certain vitamins in my food. These only could be had at low cost from liver. I hated the taste of liver. Mom experimented with it until she actually succeeded in making it something I would eat. This involved cooking the liver with bacon and various spices to disguise the taste. Dad actually liked liver. He wondered why I turned up my nose at it.

When liver was on the menu Mom always proclaimed how delicious it was and how much she liked it. Eventually, as the family prospered, and more expensive foods became available, liver disappeared. In those earlier years Mom was being sly to get me to eat it. A very long time later she sheepishly admitted that, like me, she hated liver. It should not have been a surprise when she told me. After all, Mom did give that liver-tastes-bad gene to me.

In the opinion of many connoisseurs there are only two really great European cuisines: French and Sicilian. Mom was Sicilian. Her parents were born on that island. Mom claimed she did not learn to cook at home. Maybe, maybe not. I suspect that she osmosed her mother's secrets, for Mom produced meals that were far better than any I have had in the fanciest Italian restaurants. And I do remember how wonderful my grandmother's meals were.

Wherever she learned to cook, Mom really had the gift. I suspect it was a combination of inquisitive creativity and willingness to experiment, and even to fail until she got it right. She was always trying different combinations of ingredients and amounts, finding out what worked and what didn't before she settled on a recipe of her own. But more than that, she simply had exquisitely refined taste. She knew when things were right and when they weren't.

Let's start with her chuckwagon casserole. This is a concoction of ground beef and odds and ends stirred together. Presumably something like this is what cowboys used to eat from their chuck wagons while herding cows out on the prairie.

I remember the day that Mom was inspired by a chuckwagon recipe in a back page of the newspaper. But this, like most published recipes for chuckwagon, had a Mexican flavor with hot chilies added. Mom had a different idea. Hot Mexican was not to her taste. She would conjure up her own recipe, Sicilian style. What she created was much more akin to the wonderful tomato-based pasta meals that she prepared. When perfected it was only a very slight touch less exalted than her spaghetti and meatballs.

Mom experimented with this new dish. Her major innovation was to add noodles along with a special tomato sauce. Her first attempt was good. But she wasn't satisfied. Thinking through what she had put into it she decided to try a different combination of spices. Her second attempt was significantly better — but not yet right, not quite. It took about four tries before she had what she wanted and from then on her recipe was stable. Her chuckwagon had reached perfection. Can I describe it? Of course not, you would just have to try it. Sadly, what she did is now lost to time. It can't be reproduced.

## THE TREK CONTINUES

Spaghetti dinner sounds easy. Just drop the pasta in a pot of boiling water until it softens, plop it on a plate, smoother it in tomato sauce, dump a couple of meatballs on the dish and sprinkle ground cheese over the lot. Simple! All these ingredients are readily available at your local supermarket, right? No, they are not. Not like Mom's, not anything like Mom's.

At Mom's house except for the strands of pasta and a chunk of ungrated cheese, everything was made from scratch. Everything! Not just the spaghetti dinner but her custom salad and the apple pie dessert — her magnificent apple pie dessert. There are two major secrets here and a couple of minor ones. I only know one of the minor secrets: how to make a proper salad. The rest are the result mostly of Mom's experiments and things she learned from her mother.

Start with the spaghetti sauce. She wrote out her recipe for it and gave it to me. I have long since misplaced this document. It doesn't matter. I tried following her recipe exactly. Start with tomato paste and mix in precise amounts of this and that. Cook it just this way for just this time and you will get a proper spaghetti sauce. I tried, I even experimented a bit. Nothing worked. What I got out was reasonably good but it certainly wasn't Mom's. When I confronted her about it she just said that all I needed to do was throw in handfuls of this and that and it will work out just fine. Apparently her hands were much better calibrated than my measuring cups and spoons. I gave up.

I didn't even try with the meatballs. Simple ground beef mixed with onions and various spices and some breadcrumbs. Roll the mess up, stick them in the oven for a specified time and they are ready to go. Mom's invented recipe, somewhat similar to the one she used for her wonderful meatloaf (her meatloaf had slightly different ingredients and was cooked with a bacon covering). Not the finest restaurant comes close to Mom's meatballs, or meatloaf for that matter.

Okay, the spaghetti strands are boiled for just the right number of minutes. The timing is critical, too short and the strands are chewy, too long and the strands are limp. "Al dente" as they say, is just right. But everybody knows that, so it isn't Mom's secret.

Pick a proper cheese chunk of your choice and grate it immediately before serving so it is fresh ground and has just the right taste. Different days different cheeses, to give variety to the meal.

Salad. This one I learned. Mom used an old, well-seasoned wooden bowl. Rub the bowl thoroughly with a garlic clove. Mix in the lettuce and anything else of interest that evening. Salt and pepper the mix and thoroughly toss it so that all the leaves are sprinkled with the seasoning. Pour in a good olive oil and thoroughly mix. It is important that all the ingredients be lightly coated in the olive oil. Then pour off the olive oil. Next comes a red wine vinegar. Mix thoroughly, then pour off the vinegar, leaving only a light dressing coating all that is in the bowl. The salad is ready to serve. From this instruction I developed my own way of making a Sicilian tossed salad — different from Mom's but delicious nonetheless.

Finally, dessert and apple pie. The ingredients for the crust was a standard recipe. She followed the book for that. But there was a secret that came from experiment, or from her mother. She froze the butter in the freezer so that it would crumble into small chunks when grated. These were mixed uniformly in the crust batter before rolling it out. The result was a thin flakey crust that resembled baklava but was not quite as extreme. Baklava is a Mediterranean specialty so it is likely that Mom picked the method up from grandmother.

As for the filling, there was just a taste of cinnamon, just a taste. There were other spices as well, but I don't remember them. The apples had to be fresh and crisp, of course. Put the whole thing together with the crust wrapped over the top and pop in the oven, at just the right temperature for just the right time, and out came perfection.

In all of my eighty years of life I have only tasted two apple pies that rivaled Mom's — rivaled but were just a bit less delicious than Mom's perfection. One was in western Pennsylvania Dutch country. I was twelve years old. Mom, Dad, and I were driving after dark on a country road when we came to a place advertising meals. It was in someone's old house. They had devoted their dining room to serving the traveler. There we had the most delicious meal, followed by fresh baked apple pie. The second pie to rival Mom's I savored, seated amongst working cowboys, in Leah's Café in Escalante, Utah. Don't try and find it. It is long gone.

Mom served simple meals. My top choice: salad, spaghetti and meatballs with a side of steamed vegetables, followed by apple pie. A foretaste of God's heaven. Maybe, in the not too distant future, I will once again have this heavenly meal, with Mom and Dad at the table.

# MOM'S HOME COOKING

*My dear Sarah, looking over the edge of the Grand Canyon*

# SARAH

For many years I was a bachelor. Gradually I discovered unexpected things about this ever mysterious creature I was dating. I'd fled from some women after a single date. Others became dear friends but we had different paths to follow. A couple of times I was even ready to propose but events intervened.

Then I met Sarah. All questions were immediately answered. All decisions were made. I knew from the first that Sarah was to be my life-long love. Sadly, life is sometimes too long and too short. Long for me, short for Sarah. She remains in my memory, in my stories, and in my home as a comforting spirit, always there, always just in the next room, abiding and keeping me company.

*The love of my life Sarah, and me, at our wedding*

# THE WEDDING

It is Spring. This year of heavy rains festooned the hills with glorious carpets of color. The wildflowers have gone crazy with the lavish downpours. Outside my window I see Red Tail Hawks and Peregrine Falcons carving intricate circles in the sky. It is the season of aerial mating dances. On top of the woodpile a lizard, male to be sure, pumps himself up and down to attract the attention of a female.

In nature the female usually selects a mate, not the other way around. In most cases that is likely true in human society as well, although Society has somehow thrown a monkey wrench into the process, and things have been twisted around more than a little.

Looking back at my own experience with my dear Sarah, I suspect my life with this remarkable woman was not entirely my free choice. Knowing Sarah as well as I came to over the years, I realized that most likely I was the catchee rather than the catcher. Whatever the case, it worked out splendidly.

After a loving courtship we started planning a wedding. Sarah had it all figured out: we would be wed at the Mission San Juan Capistrano. "Okay," I agreed in my innocence, not knowing what that meant. There is something about that storied Mission that couples find particularly romantic. Perhaps it was an old song, or an old movie. Whatever the reason, every year thousands apply to the Mission for a wedding. With a handful of exceptions, an almost equal number of thousands are turned down. Weddings at the Mission are reserved for Mission parishioners, and only then if they have been attending for at least a year. And we lived many miles from San Juan Capistrano.

Given the great demand from young romantics, free enterprise did find a way to cater to many who were eager to wed in this storied spot. For a few thousand dollars, quickly escalating to several tens of thousands depending on how elaborate the wedding, you can get married at San Juan Capistrano. You can get married at the town's vacation resort, that is. The day of the wedding the resort is yours. If you wish for more romance you can wander the extensive grounds and gardens of the nearby Mission; however, the lavishly decorated, but fragile, ancient chapel is off limits.

Still, Sarah was Sarah. At some gathering she had made the acquaintance of the Monsignor in charge of the mission. They instantly became best buds, and he her Confessor. When we decided we really wanted to be married there, she simply asked, and it was done.

The place settled, we found there is a lot more to do before a wedding. The Monsignor wanted to interview me to see if I was a suitable companion for his dear friend. After all, this was going to last a lifetime. A friendly chat ensued, no big deal. Until, that is, the good priest said the words of absolution. Surprise! I had just had a confession. Being a Catholic I was quite familiar with the standard protocols of confession, and this had been nothing like that. Well, I was happy to receive absolution. Onwards.

The next thing was to get a marriage license. Sarah and I trekked off to Santa Ana to take care of the appropriate

## THE WEDDING

documents. After signing the papers and paying the fee we turned to leave. I had a question, though. Pointing to the clerk's window, I asked the nice young lady, "What are those crepe paper ornaments hung all around your window?"

"Oh, didn't you know? You just got married!"

Silly me, I had thought that a civil wedding required a judge and ceremony, or something.

Whatever the civil authorities may have believed, we were not officially married until our ceremony at the fabled Mission San Juan Capistrano.

*Sarah, my amazing bride, and my wonderful sister Marie, at our wedding*

# THE WEDDING, PART TWO — DIPLOMACY

The day of our wedding I supported my very nervous bride as she tottered her way up the steps to the altar. Traditionally it is supposed to be the groom who is a bundle of nerves, but I felt light as a feather with the joy of the day.

It was not a large crowd in attendance, just a few friends and immediate family — Sarah's young son, my parents and sister, and my favorite Aunt Fran and Uncle Nick who had flown in from New York for the ceremony. Uncle Nick, he's the one who is a Papal Knight and was, at the time, a major power on Wall Street.

Sarah and I were blessed to have other sparkling talents as our friends — a highly decorated soldier, a senior executive of the aerospace industry, a film director and department chairman of a major college, a very distinguished scientist, a foreign language newspaper publisher, a noted and popular astrologer, sundry gifted musicians, artists and writers whose great successes still lay in the future — and, of course, equally interesting wives and husbands. Many of those in attendance were also multilingual. And thereby hangs a tale.

After the wedding we adjourned to a nearby gourmet restaurant for the wedding supper. The group was distributed among several tables, with the high table being reserved for our immediate family. After a delicious main course Sarah and I split up and started circulating among the guest tables, spending substantial time at each. The wit that was flowing across the white linen of the circular tables was fun, and I was glad our guests were enjoying themselves. Nonetheless, underneath the laughter something seemed wrong.

At first I couldn't place what was troubling me. Then I noticed that several people had become rather cold towards the waiters — three tall, handsome, young French men. Very polite and capable servers who spoke excellent, slightly accented English. But while they were exceptionally polite when waiting on us, on the sidelines their French conversation had quite a different tone. They clustered near the center of the high table, readily available for any call, as one would expect. But occasionally a smile, almost a smirk, would cross their faces and a slight snickering could be heard as they compared notes on the wedding party.

When the tension in the room had risen to a palpable level, my sister Marie found the solution. Highly educated and fluent in several languages, Marie also has a gift for finding subtle solutions to intractable problems. Perhaps that is why she has since had such a successful career as a Foreign Service Diplomat with the U.S. State Department.

Still at the head table, Marie motioned one of the nearby waiters over. He came and bent down low, for my sister deliberately kept her voice very soft. The other waiters, having nothing better to do, were watching and listening, as were several of the guests. Marie simply requested another glass of wine. She asked for it, though, in flawless Parisian French. The waiter shot bolt upright, his face flushing pink with embarrassment. The other waiters, observing the transaction, had a similar reaction.

## THE WEDDING, PART TWO — DIPLOMACY

Several people around the room had been watching to see how Marie would handle the situation. A wave of quiet laughter, and knowing glances at the waiters, spread quickly around the assembly as all became acquainted with what had just transpired. The pink deepened to scarlet. Problem solved.

*Author Chester L. Richards holding his and his wife Sarah's amazing dog, Hector*

# HECTOR

HECTOR WAS A PUP, a Golden Retriever pup. He was just old enough to start his primary training. So we took him to doggy school. Two schools, in fact. The first was tutored by a nice lady who was convinced that she had a psychic connection with her pupils. That may have been, but Hector did not much respect this teacher, so we found another.

Elaine, teacher and owner of school two, was a crusty, no-nonsense lady who very successfully specialized in training German Shepherds for show. She took one look at Hector, as he bounced around the training ground, happily greeting other doggy students, and firmly declared: "This dog will take a First at Westminster!"

Now, at the time, I had no idea what Westminster — the pinnacle of dog shows — was. I learned about it later. I also learned that no Golden Retriever had ever taken a First there; the very idea was anathema to the judging gentry.

Sarah had picked Hector out of the litter. It took only a few seconds for her to make the decision. She later explained that not only was he extraordinarily handsome (so were his siblings); of much more importance, he was the pup with the lively curiosity and boundless energy that signaled great intelligence.

Hector very much liked Elaine and learned quickly from her. And, of course, the respect was mutual. In the beginning there was only one thing about Hector that had concerned Elaine. On the first day of class the lad bounced up to a young Rottweiler to say hello. In response the Rottweiler said something to Hector that made our Golden flip over on his back in submission. This would not do for a dog that was destined to be a Champion. Elaine quickly put a stop to such behavior.

Did I mention that despite his momentary capitulation to another dog, Hector was headstrong? For weeks we all struggled to teach him the fundamentals. Actually, being very smart, Hector picked up the lessons in an instant. He fully understood what was being asked, always (or almost always) obeying Elaine right away when she took hold of him. But with us, Hector chose to go his own way, visibly laughing at our struggles to get him to obey. Except, every once in a while, and just for fun, he would do everything perfectly. And, every once in a while, just to show who was boss, he would totally ignore Elaine's commands and then spin around and look up at her — his face radiating joy. At this Elaine would start to explode, but the fury invariably sputtered into laughter at Hector's indomitable good spirit.

Hector certainly was a charmer. As we drove to and from the training ground every week, he would toss his head and grin at the other cars. The response was invariably honking horns and gleeful waves as laughing people passed us. At the training ground, the other dog lovers instinctively gathered around Hector. He, of course, basked in all the attention.

Still, as time passed and Hector failed to fall fully in line, Elaine got progressively more frustrated. Eventually the time came when Elaine, Sarah and I simply gave up on weekly training. The solution, we all agreed, was to send Hector to

# HECTOR

Boot Camp. There he would have the nonsense knocked out of him. Elaine promised that Hector would return a well-disciplined veteran, an expert with advanced training. So, one day we trucked Hector up to Elaine's place, far out in the country.

Elaine's immaculate establishment had separate kennels for the several very handsome German Shepherds that she was preparing for show. Hector quickly made the acquaintance of these wonder dogs and settled right in. Satisfied with the arrangement, Sarah and I started the long drive home. We would not again see Hector until some weeks had passed.

Periodically we phoned Elaine. She reported that Hector felt right at home and was making good progress. He was proving to be everything that Elaine had hoped for. We started talking seriously about his career as a show dog.

The day came when Elaine informed us that Hector was ready to go home. We drove the long journey over to Elaine's place. Hector, catching sight of Sarah and myself, bounded up to greet us with great gobs of slobber.

Elaine put Hector through his newly learned paces. Sit! Lie Down! Stay! Come! And much, much more. Hector responded perfectly, his head always cocked slightly towards Elaine to anticipate which command would come next. Then it was our turn. Again, Hector responded instantly to every cue and command. Elaine was pleased. We were impressed. Hector was visibly proud of his new skills — strutting around the grounds as if he owned the place.

But Hector's new skill in obedience was not the only change in him. Association with the friendly Huns had converted Hector's happy-go-lucky bark into a deep throated explosion. By the sound of him, happy Hector was now an honorary member of the German Shepherd Society. We would soon learn he was a member in more ways than one.

Realizing he was now free to leave Boot Camp, Hector dashed up to our car, waited for the door to open, and hopped right in. We three were on our way home.

The next week we drove Hector down to the old training grounds. Elaine wanted to show the other folks what could be accomplished with even the most headstrong student. When we arrived Hector caught sight of his old nemesis, the Rottweiler. Immediately he ran up to greet that stocky dog. After a brief conversation, the Rott suddenly flipped over on his back, legs up in submission. How things had changed since Hector had become a Hun! Elaine moved quickly to put a stop to this new behavior. A show dog simply must not intimidate other dogs — at least not so the judges would notice.

Then it was time for Elaine to put Hector through his paces. First she lectured the group on her methods and how well they had worked even with the notoriously headstrong Hector. Next Elaine commanded Hector to sit. Hector sat, cocking his head towards his leader, alert for the next command. Elaine told Hector to lie down. Hector went down, as ordered. Elaine directed Hector to stay. Then she walked away. Hector stayed. Elaine told Hector to come. Hector, his face breaking into a great grin, just lay there, alert but unmoving. Once again Elaine commanded Hector to come. Once again Hector, now having gained the upper hand, remained where he was, the heels of his paws dug firmly into the ground to show that he was not about to move. This grew into a considerable contest until Elaine, embarrassed by Hector's intransigence, finally gave up and turned away in disgust.

# HECTOR

Whereupon, to the delight of the crowd, Hector promptly performed all the commands in precisely the sequence they had been given. The group dissolved in laughter. Elaine, red faced, just sighed and shrugged her shoulders. Hector was going to be a real challenge.

If Elaine was a good teacher, an even better one was Sunny. Now, Sunny was a little seven pound, orange tabby micro-cat. And she was definitely the boss in our house. Right from the start Sunny had Hector under control, just as she dominated the rest of the neighborhood.

Next door lived two large, identical, twin cats. Before we moved in, our yard had been part of their territory. This changed the first time these beefy cats encountered our diminutive Sunny. What a racket! Chinese Opera, Sarah called it. And Sunny was quite the singer. Lots of stylized posturing and standing hair, ears flat back. Sunny gave no ground. She slowly advanced. The Twins, equally slowly, gave way. Suddenly they skedaddled, Sunny racing after them and nipping at their tails. It was clear who was in charge.

The Twins soon became Sunny's best buds. They would come to visit and practice their Chinese Opera, after which all three would pal around our yard. The Twins disappeared after a year or so. We assumed that coyotes had gotten them, since that was their most likely fate. Their real fate was far worse. Years later their mummies were discovered in the attic of a recently reroofed house — the Twins had been trapped there by their fear of the workman as the roof was sealed. Not all kitty hazards have teeth or claws.

Sunny had another occasional visitor, a big cat we called Scar from the deep wounds that had been gouged across her face. Scar, too, became part of Sunny's entourage, as did several other cats in the neighborhood. Scar was feral, but felt sufficiently at home with us that, after some years had passed, as she was growing old and slowing down, she dropped her kittens in our wood pile. This happened just days before a coyote devoured her. (We saved these shy wild kittens, tamed them, and adopted them out.)

Sunny tolerated Hector the wee puppy, then grew fond of him as he grew up. But Sunny was not *about* to tolerate one particular behavior. We had been paper training Hector prior to getting him fully house broken. But one day, bladder much too full, he let loose in the living room. It didn't hurt the room any, for we had had it refloored with Mexican pavers. Though Hector had not damaged the floor, meticulous Sunny was having none of it. No dog of hers was going to piddle a puddle in the middle of *her* living room. Sunny raced across the room, swatted Hector firmly across the muzzle and led him to the nearest door. After which Hector, when he needed to do his duty, would walk to that door and gaze fondly at the handle. Hector had instantly been fully house broken by Mistress Sunny.

As you might imagine, Sunny was a feisty, fiery little feline. But her character was tempered by wit and humor. She was always fond of practical jokes. One day I heard Sunny's plaintive cries from the patio. She was trapped up in the shading overhead lattice, barely hanging on for dear life. I hurried to haul out a step ladder before she fell and injured herself, but when I climbed up the ladder and looked for her she was gone. At my feet I heard a chuckling noise and looked down. There was Sunny, grinning up at me. Gotcha!

And, of course, there was also her favorite fright gag. Hearing us coming she would hide around a corner, then pop out at the last moment, walking upright on her hind legs, her arms splayed out in front, looking for all the world like a miniature Frankenstein's monster. *Boo!*

Sunny had a tender side as well. One morning — on Sarah's birthday — we were sitting at the breakfast table, as usual. Sarah was delighting in the presents I was giving her. Sunny, observing Sarah's joy, ran outside. A couple of minutes later she returned, jumped up on the table and presented Sarah with her birthday gift — a frightened little bird that was carefully nestled in Sunny's mouth. We let the bird fly off, followed by much praise for loving Sunny and her gift.

As loving and funny as Sunny usually was, she could be truly scary at times — to those who deserved it. One day Sunny disappeared. She did not come up from the wilderness below our house when called for dinner. She did not show up the next morning, or in the days that followed. Gradually we gave up hope. Living at the edge of the wild has its charms, but it has predators, as well. Coyotes are in abundance. So are hawks, occasional eagles, great owls and even mountain lions. Then, too, there are the rattlesnakes to nip the unwary. Most domestic cats do not last long in this environment. Sunny had lasted longer than most.

Four days had elapsed since Sunny's disappearance. Sarah and I were having breakfast, Sarah reminiscing about Sunny: "I can just see Sunny poking her head up above the edge of the patio like she always used to do." The *instant* Sarah finished saying that, Sunny's head popped up above the edge of the patio. My belly did a flip-flop! This was weird.

Sunny was beat up and bedraggled. We cleaned her up a bit and rushed her down to the vet. The doctor cleaned her some more then reported what he had found. All across her abdomen and back were deep claw marks. Sunny had been carried away by some big bird — a Red Tail Hawk was the likely culprit. Much to the surprise of that fearsome predator, Sunny fought her way free and must have fallen far. It took her days to recover enough strength to make her way the long distance home. No doubt that bird would never forget its painful encounter with our fierce, scary Sunny.

So Sarah and I had two companions, Sunny and still growing Hector. That was about to change.

One early morning, as usual, I opened the service porch to let Hector out into the backyard to run. There I found a scene of horror. Hector was lying in a scum of urine and feces. In his agony he had scattered the mess far and wide. He looked up at me with sad eyes and tried to rise, but that was beyond his strength and he laid his head back down. A few minutes later a doctor from the local veterinary hospital arrived to bundle Hector up and take him away. In tears I cleaned up the mess and waited for the word about Hector. This eventually came. Hector had been poisoned by ethylene glycol — common antifreeze — which someone had put in his outside water dish. Because of its sweet flavor it is almost impossible for an animal to resist and Hector had nearly drunk his dish dry before bedding down for the night.

The vets held out some hope that they could save him. Hector was fighting hard, they said, and it might just be enough. But the all-day efforts of the team of doctors were to no avail and young Hector died late in the afternoon. The stress and shock at the loss of this popular creature was so great that the head of the medical team took a leave of absence and headed for the high mountains to regain his composure. He was gone for weeks.

# HECTOR

It was pretty obvious who did it. We had a neighbor who truly hated Hector. It started simply enough with Hector's usual joyful greeting of a couple of barks, hello, then quiet after that. Unlike all other neighbors who were happily greeted this way, and said hello in return, this particular neighbor immediately started yelling at Hector to shut up. Hector quickly assessed the character of this woman and responded accordingly. Even when Hector was quietly minding his own business the mere sight of him was enough to trigger the woman into a screaming rage, whereupon, of course, Hector responded accordingly.

One day the woman appeared at our front door. I recoiled from the whiskey on her breath. She slurred out, angrily, that she was trying to sell her house and that Hector was preventing the sale. If we didn't get rid of that dog she would get rid of him for us. Sarah calmly offered to bring Hector inside whenever she wanted to show her house. All she had to do was ask. The woman was having none of it and she stomped off with a final warning.

A few days later our gardener was briefly in the backyard. This was a gardener that we shared with just this particular neighbor, and no others. The next day Hector was dead.

The sheriff listened to our tale and agreed that the woman had probably hired the gardener to poison Hector. Of course it was a felony, and a serious one at that, but one that almost certainly would fail in prosecution. The sheriff advised us to drop the matter, which we did, except for firing the gardener.

A few weeks later the neighbor sold her house and moved out. The neighborhood was much improved after that.

# KITTENS

A FEW YEARS PASSED AFTER WE LOST HECTOR. I was now self-employed as an engineering consultant. I was making ends meet, but just barely and with very long hours. Sarah had recently suffered an automobile accident which left her crippled and in permanent pain.

No longer able to sing professionally nor practice as a psychotherapist, Sarah had time to reflect on what would rebuild my morale and hers. She did some research and came up with a solution.

We needed a couple of kittens. Not just any kittens, either. These new additions to our family had to be Bengals. Bengal Cats are a new hybrid breed formed by crossing short haired domestics with wild Asian Leopard Cats (P. Bengalensis). Through a careful genetic dance of cross breeding and in-breeding, Bengals turn out to be gentle, friendly, energetic, highly intelligent and gorgeous. And, they retain all the hunting skills of their wild ancestors. And so it all proved to be with our new kittens.

The kittens' daddy was a brute. Bobcat sized, he sat at the far corner of a large cage, glaring at me. I walked up and said, quietly, "Hello there." There must have been something in my tone of voice. The big cat instantly leaped across the cage onto a nearby platform, his eyes inches from mine. He bared his teeth and roared! I recoiled. "He really likes you," the breeder laughed. I had my doubts. "My boy seldom greets in that way," he explained, "and then only to good friends. You've won him over."

The mother was very special. She had the long pointy face and rosette spots of her wild ancestors — very different in appearance from domestic cats. But in a more important respect it was obvious that this was no ordinary cat, even among Bengals. The mother's coat was not fur, it seemed. Catching the sun as she glided across the room, her pelt appeared composed of infinitely fine strands of shining gold. This great beauty *flowed*, and reflected, like liquid metal.

They call it *glitter*, and hers was far more beautiful — more metallic — than any I have seen since. Glitter is unique to Bengals. It is caused by microscopic bubbles of air in the hairs. No one knows *why* it occurs. No one can predict *when* it will occur. No one knows *how* the glow will manifest itself, until it does. It is one of those chance treasures, and mysteries, hidden in the hybridized genes.

We came home with two kittens from the litter. Sarah selected the little male. In the time immemorial way that all females have, one of the kittens seduced me.

This female was the last born and had not yet been weaned. From the pet store I obtained a special potion to help the kitten transition away from her mother's milk. The kitten sniffed the concoction, wrinkled her nose in disgust, and proceeded to gobble up the hamburger we had put down for her brother. So much for weaning.

What to call these new additions to our family? I tried out various names, but Sarah shushed me. "Just wait a bit," she advised. "When they are ready they will tell you their names."

## THE TREK CONTINUES

Soon our Mink became Merlin and Misty our snow. Merlin's coat not only had the color of fine mink, it also had the soft, luxurious texture of that animal. Misty had dark gray spots on a snow white base. Later, as she grew into adulthood, Misty developed a long, silky, silvery-gray overcoat. She ended up looking like a Russian Blue, but with the spotted white undercoat still remaining. Both cats were strikingly beautiful.

*Merlin (the first Merlin) is the mink in front, in the white round cat bed. Misty is the blue cat behind Merlin, lying on the carpet next to Misty and the bed.*

KITTENS ARE A DELIGHT — a tumbling, bumbling, concatenation of bumptiousness and shyness. As these two slowly explored out into the house from their safe home base near the kitchen, distinct differences in their personalities began to appear. Merlin was filled with the boldest curiosity. Misty, by way of contrast, was reserved and observant. Misty was quite content to let Merlin take the lead in their explorations. But she watched carefully and joined Merlin once she was convinced that it was safe.

Misty was very athletic. She invariable bested Merlin in their playful combats — hissing, growling, squalling, wrestling affairs those squabbles were. The two would go at it for some minutes, then, quite exhausted with their fun, the two would tenderly groom each other and collapse, lovingly embracing, into a snoring kitty pile. These squabble fests became regular morning and evening rituals and continued for a lifetime.

There was, for a time, a remarkable variation in their mock combats. Having taken their places at opposite corners of the room, at some signal the two would rush at each other, leap high into the air and collide, then tumble to the floor, already growling and wrestling as they fell. After a few months these aerial gyrations reverted back to just the usual floor exercises.

Misty, in time, grew to be a great athlete. Merlin, on the other hand, was the more intelligent of the two. Indeed, Merlin was a kind of kitty genius. Merlin's great intellect was first manifest when the kittens had been with us only a couple of weeks. By this time the two had explored the kitchen and the dining room. They had not yet penetrated into the living room, and though they could see the entry area from the kitchen, the invisible hall beyond was still a distant mystery.

One memorable day the two kittens were playing in the kitchen when a large raccoon came lumbering by outside the dining room. Without hesitation Merlin rushed into the still unexplored living room and jumped up onto the windowsill to get a better view. The raccoon continued waddling past, then disappeared around the chimney which blocked the view down the rest of the house. Merlin sat there for a few moments. I could see the wheels turning as he thought this thing through. Suddenly, and signaled by a startled expression on his face, the light bulb flashed on. Without hesitation Merlin dashed to the hallway, ran pell-mell down the long hall, darted into the bedroom and leapt upon the bedroom's windowsill, there to watch, once again, the raccoon ramble by.

## THE TREK CONTINUES

How had this tiny kitten constructed, in his imagination, the plan of the hidden parts of the house, how had he figured out where the raccoon had to be going, and how had he found the gumption to throw caution to the winds and instantly act upon his flash of insight? Kitty genius indeed! Bold kitty! But that was our Merlin.

If Merlin had the brains, Misty had the brawn. Fully grown, but still young, Misty was sitting on the floor of the dining room. Eight feet above a fly bounced against the ceiling. Misty's head swiveled back and forth as she tracked this annoying intruder through its random flight. Suddenly, from her seated position, Misty leaped up and grabbed the fly as it bounced along the ceiling. Back on the floor, and with a very satisfied expression on her face, Misty tongued the fly to her teeth. Crunch! The fly met its swift demise.

As I said, Misty was a great athlete. If there had been a Kitty Olympics Misty would easily have taken the Gold.

Kittens grow quickly. Soon it was time for them to be introduced to the great outdoors, under supervision, of course. Near the house Sarah and I supervised. Later, as the cats became accustomed to the outside world, Sunny took over their training, leading them into the wilderness, teaching them how to survive the predatory hazards that abound there.

As expected, the kittens made mistakes. One day I was uncovering the swimming pool when Merlin hopped onto the free remnant of the cover and started strutting around, amused by its oscillations beneath his feet. I tried to warn him off that dangerous place, but Merlin ignored me and walked ever closer to the cover's drooping edge. Suddenly, he started sliding. Frantic, he tried clawing into the material, but the fabric resisted and he slid, backwards, right into the deep water.

# KITTENS

I was ready to jump in, clothes and all, to rescue the drowning cat. Except, this cat was not drowning. Not a bit! Merlin looked at me with an expression of pure joy, and started paddling around the pool as if he had been born to it. Now, I had long been aware of a Bengal's affinity for water. This breed loves to play in the tub with a couple of inches of water to splash around in. But swimming in a pool was not even rumored. Yet, there Merlin was, happily swimming around without the slightest sign of distress. After a while he found the steps, climbed out of the pool, shook off sprays of water and came trotting up to soak me, as well.

Though Merlin never showed fear of the water, Misty, by way of contrast, kept her distance from the pool. Merlin often quite casually walked along the very edge of the pool, never concerned about falling in. Once in a while, when he was thirsty, he would lean far down and take a drink, much as do the larger wild animals around here. Of course, as I had discovered during a close examination of his paws, Merlin had webbed feet. Maybe an ancestor of Merlin and Misty had acquired genes from an Asian Fishing Cat — a wild cat that dives deep in lakes and rivers to catch the fish that it feeds upon. Such cross breeding would explain the very large size of our kittens' dad, as well as Merlin's webbed feet.

Merlin was remarkable in another way. He did not meow, like most cats. His voice was a wonderful musical instrument. It had the resonances of a bassoon's upper register. His greeting was a fragment of a symphony. Charming! To this day I still hear Merlin's voice when a symphony orchestra is playing.

For the fast-growing kittens the wilderness proved to be fun. The wilderness was also a happy hunting ground, part of the fun, no doubt.

## THE TREK CONTINUES

Their greatest delight was chasing squirrels. Now, squirrels love trees. They feel nice and safe high up within the protective arboreal thicket. Oddly enough, so, too, do Bengals. Perhaps this is the legacy of another of their wild ancestors, the Leopard cats who hunt in the trees of Asia. In any case, Merlin and Misty had good sport merrily chasing the squirrels right into their lairs deep among the branches. "Okay," a squirrel would respond, "see how you like this move." Whereupon the squirrel would dash right out to the bendy, whippy end of the limb and leap to another, many feet away. Misty, in the lead with Merlin close behind, would chase after the squirrel, but not quite so far out on the branch since Misty was heavier than the squirrel. Much to the squirrel's astonishment, and evident dismay, Misty, too, would make the long leap to the other branch, Merlin leaping after her. Don't get me wrong, this was pure play for the fast-growing kittens — nothing malicious intended. It took some time for the squirrels to recognize this and relax — somewhat. Still, the squirrels never let Misty and Merlin get *too* close.

The serious hunting was done on the ground. At the time the neighborhood was suffering from major infestations of gophers and their predators, rattlesnakes. Misty and Merlin took care of the problem. First they went after the gophers. With their primary food supply gone most of the rattlers subsequently went missing. The few, hunger weakened, vipers that were left were easily dispatched by these lightening quick cats.

Hunting gophers required precise coordination. First, the kittens used their keen hearing to locate the two openings to each gopher den. While Merlin attacked the entrance with furious digging Misty waited patiently at the exit to catch and dispatch the panicked gophers as they fled from the security of their nests. The lawns in the neighborhood were soon blessedly free from their myriad gopher holes.

There was something uncanny about the way these two worked together. On occasion I would see them walking side by side, in perfect step, their ears swiveling in unison, their heads swinging back and forth as one, their tails switching side to side, together. The two looked for all the world like a single organism with eight legs. Sarah maintained that, in that state, they really were one for they were communicating telepathically.

Sarah's theory was that these higher level animals communicate by exchanging pictures rather than words. Sarah apparently could do some of this herself for I often watched her easily taming wild animals. She said she simply told them that she was one of them. They evidently believed her. There were incidents from her past which made her theory at least somewhat plausible: One summer she worked as a Forest Ranger watching for fires from the top of a mountain. Living near the lookout tower was a bear. This momma bear, together with her cub, would, every week, walk close by Sarah's side as she made the long trek down to the base camp to pick up supplies. Sarah's affinity for wild animals attracted the attention of Ella, the Shaman of the Nez Perce Indian tribe. Ella trained Sarah to be a tribal shaman and then adopted her. That was how Sarah became a member of the tribe. Having learned her lessons well, Sarah said that Ella knew far more about wild animals and how they communicate than any academic biologist.

Whether Merlin and Misty's communication was telepathic or by some other means, it was fascinating to watch them at work, or at play, always strangely synchronized.

As Merlin grew towards adulthood he became increasingly strong-minded. When the call to dinner went out, both cats would come running up the hillside from below, sail high over the fence and stampede their way into the kitchen. Gradually Merlin started lagging far behind Misty at the dinner call. On occasion the best part of an hour would elapse before he would come trotting in. The day finally came when Merlin no longer responded at all. It seemed obvious that Merlin was going feral. That evening I went searching for him and found him far away. We permanently grounded Merlin, and the other cats with him. We still let them outside, on occasion, but only in the backyard and only with supervision.

It was during one of those backyard excursions when something happened to the still young Merlin that changed the lives of all three cats. Merlin had often jumped from the backyard fence onto the roof. It wasn't that far a leap for these Bengals, although most cats would not make the attempt. On this particular afternoon Merlin hesitated surprisingly long before making his usual jump. He jumped. He missed — clinging briefly to the edge of the roof. Down he fell to crash hard on the pavement below. No fancy cat trick of flipping over and landing on his feet, he landed flat on his side.

We rushed Merlin down to the vets. The doctor there x-rayed him. The news was unexpected. The fall had done no more than bruise Merlin and wound his dignity. The x-rays showed something else, something much more serious. Merlin was suffering from cancer. The advance of this disease probably accounted for Merlin's failed leap. Further diagnosis

# KITTENS

indicated that the cancer was Spindle Cell. This was bad news indeed, for this type of cancer usually kills within weeks or months. The vet held out only slim hope — there was one recorded instance of a spindle cell victim reaching the age of five years. This hope was very slim, but you never knew. Perhaps the young Merlin would survive longer than most.

For a while, at the doctor's suggestion, we tranquilized Merlin. This spirited cat fought the pills, secreting them in the pouch of his cheek and spitting them out later. He did not like the torpor they induced. Neither did we. After some months we discontinued them. Better that Merlin live a lively short life than live a long listless one.

Merlin lived. And he lived some more. And still more. Our visits to the vet became less frequent until they settled into a routine six month interval. The years passed. The cancer never went into remission, but it no longer was growing. After Merlin's fifth birthday all the vets at the hospital started calling Merlin "The Miracle Kitty." That was how he was always greeted in the years that followed.

Sunny grew old, arthritic and cranky. She would no longer have anything to do with our Bengals. She still enjoyed a good pet, but otherwise kept to herself, off in the corner. By then I had given up consulting and was working a regular job (my hours had, by this choice, been reduced from more than sixty per week down to the usual forty, and my income had substantially increased). One day I returned home to find that Sunny was missing. Sarah had taken her down to the vet's and had her put down. She had not told me that she was going to. She wanted to spare me the anguish that always accompanies the last moments with a much loved friend.

## THE TREK CONTINUES

As he grew and reached the age of young Hector's demise, Merlin's behavior changed. There was something familiar about it, but something that I could not quite put my finger on: Something about the way he carried himself. Something about the way he tossed his head. Something about the way he occasionally crouched, a big silly grin on his face. Slowly it dawned on me that this cat had something of old Hector in him. The more I observed these un-cat-like behaviors the more convinced I became that maybe Merlin had a piece of Hector's ghost inside of him. He even *barked* at the squirrels and birds — not what you would expect of a cat — cats chitter, they don't bark.

Merlin for sure was very much a cat. Echoes of Hector were a minor part of his personality. Still, I could not help but wonder, perhaps the tiny kitten's brilliant analysis of the unexplored parts of the house had a touch of someone else's memory to it. Perhaps Hector was revisiting us in an odd sort of way. Sarah noticed it, as well. We frequently talked about it. Whatever the cause, coincidence or not, there was a spooky similarity to their personalities. Thank God! I still very much missed Hector. Even after the passage of years the mystery still occasionally returns to haunt me.

The years passed. In Merlin's tenth year I underwent a difficult surgery which kept me home for some weeks. One evening, while I was recuperating, Misty approached Merlin for their usual wrestle. As usual, Misty started with a Chinese Opera. But there was something in Merlin's responding voice that was different. Misty approached, ready to begin the melee. Merlin backed away. Strange. Misty was puzzled. Again she approached, but again Merlin backed away, growling a warning. Misty pounced and Merlin let out a howl of pain.

Misty backed off and just sat, puzzled and concerned. Carefully I examined Merlin, he showed small signs of distress but otherwise everything seemed normal. The next day we took Merlin down to the see the vet. Overnight Merlin's cancer had exploded into a large tumor swelling out from his side. The vet gave us some morphine for Merlin. He had, at most, a few weeks to live.

The morphine helped. We set Merlin up in a room of his own. Misty could visit him, but only under supervision. Misty took special care of Merlin during these few days. Patiently she groomed him. Patiently, lovingly, she comforted him. It helped. Merlin's appetite returned and so did his morale.

A few days later he once again stopped eating. Again we took Merlin to the vet. The vet increased Merlin's dose and we returned home. As miserable as I was from the pain of my surgery, it was obvious that Merlin was worse off. That evening, with this increased dose of morphine, Merlin again was able to eat a bit of supper. But his strength was fast failing. I visited him frequently. As I opened the door he would rise to greet me and sing with his musical voice. Then he would sink back down, his strength depleted.

By midnight his hind quarters were paralyzed. I spent most of the night with him. We talked, we sang songs — kitty songs that Sarah had composed. Merlin played his bassoon for me, between snippets of purring.

With the sun rising I called the vet and made the arrangements. Sarah and I took him down. Tenderly I cradled his head in my hands as the poison took effect. His eyes glazed still. He was gone. We cried. The vet cried. The staff cried. The Miracle Kitty was no more.

## THE TREK CONTINUES

Misty took it hard. For days she wandered the house, yowling, calling for Merlin to return. Maybe he was hiding somewhere. Misty kept going back to where she had last seen her very sick brother. Carefully she examined every part of the room, even squeezing behind the furniture, looking for where he might be hiding. To no avail. For weeks, then months, Misty searched for Merlin. But he was never there.

The day Merlin died Hector would have been seventeen years old and ready for his grave. Hector finally left with Merlin.

These small creatures become family. Each has its own unique personality. Each has its own charm. Each becomes so much a part of one that the inevitable loss of their short lives becomes increasingly hard to take. Even so, it is hard to not adopt someone new when someone old vanishes. One day, months after Merlin died, on the drive home from work I found myself crying out repeatedly, "I want my Merlin back!" Funny, Sarah responded, when I told her this, she had been saying the same thing all day long.

That Saturday we drove to a breeder's establishment and selected two new Bengals. Sarah selected a cat who had Merlin's face. He became our new Merlin. I asked the breeder if she had any Marbles. A marbled Bengal is relatively rare and very beautiful. The breeder brought me a small ball of fur and placed it in my arms. This tiny creature snuggled into the crook of my arm, looked up at me with an intensely serious expression and said, with a powerful voice: "Yeeeeooooowwwww!" "Take me home Daddy." We took him home. This bundle of energy became our Sparks.

# KITTENS

*Sparks*

ONCE AGAIN OUR HOUSE WAS FILLED with the explosive delight of rambunctious kittens. The kittens created their usual magic — Sarah and I were young again. For a time we were intensely happy. This despite Sarah's growing incapacity and pain.

Young Merlin and Sparks were not yet fully grown when the worst happened.

One day I came home a bit late from work. When I arrived home I did not hear Sarah's usual cheery greeting. Something was wrong. She was not where she normally stayed. I finally found her slumped in her wheelchair outside the bathroom, waiting for me. She said she had been there a long time. I told her why I was late. "That's okay," she said — her final words to me. She lifted herself, took a few steps into the bathroom and collapsed — her face gone gray, her pulse not to be found. My

## THE TREK CONTINUES

beloved Sarah died in my arms as I frantically waited for the paramedics to arrive.

If this has happened to you I need say no more, you will know. If not, no words can tell of the anguish. You simply cannot know.

As I soon discovered, there is among us a very special community. It is not normally remarked upon so I did not know of it, but it is there, ready to help when needed. It is not an association that you will want to join. But it is there and you might someday discover that you are a member. As word of Sarah's death got around, people started coming to me — people I had never met before. These were not ordinary people, these were those who had suffered a similar loss. Instinctively they knew my state of mind and came to help. And help they did.

So now we were four — myself and three cats. We all suffered. We all changed. The personality of the kittens became sharply different. Merlin withdrew. No longer able to be Sarah's constant, joyous companion, he stopped purring for a year or two. Sparks became more aggressive, in his confusion playing pranks when it was not appropriate. And Misty was hit the hardest of the three. She had, by now, seen too much death and she knew exactly what it meant. She could no longer tolerate Sparks' jolly japes. She could no longer give as good as she got. Now, with the added stress, she declined into a nervous breakdown. I had to isolate her.

And then my very old father died. I took Misty across the continent to live with my mother. Misty, who had never been more than a couple miles from our home, turned out to be a great traveler. She charmed the security personnel at the airport. She charmed the airport attendants. She charmed the flight crew of the jet. She charmed her fellow passengers. And she charmed and consoled my mother in her grief. The two of them became inseparable.

The years have passed. Merlin and Sparks helped me heal. We three would sit together, Sparks in my lap, purring. Shy Merlin curled up beside me, purring likewise. Like the Magi of old we three philosophers would sit by the hearth fire and ponder our place in the world. Three's a good number.

*Merlin bathing Sparks*

*Sarah with our cat Sunny*

## MOMENTS WITH SARAH

"Hi, Chet, welcome home. How was your day?" That greeting enveloped me in warmth and love every evening when I walked in the door. The lilt in her voice swept away the fatigues of a long workday. It was always good to be home.

Now, though Sarah is long gone from this life, somehow I still feel her presence. It's as if she is nearby and will appear at any moment. I know that cannot be. I must be content with sweet memories of our travels together through life.

# MUSICAL GIFTS

AMONG HER MANY TALENTS SARAH was a professional singer. She had sung with the Seattle Opera and she had teamed up with the young baritone Thomas Hampson, the future operatic superstar, to sing duets at weddings and other functions. The two of them met when they were under the tutelage of the same voice coach in Spokane.

Other than a fine spinto soprano voice, Sarah had two other rare musical assets: perfect pitch and an eidetic musical memory.

Perfect pitch was a blessing and a curse. This talent gave her work as the pitch anchor for various Southern California choirs. From the audience point of view she would stand in the upper left corner where her voice could lock-in the pitches of the rest of the choir. The curse came when we were watching famous divas on TV sliding into the notes and never quite making it. It drove her nuts and we would have to change the channel.

Eidetic musical memory is the rare gift of being able to permanently remember a piece, exactly as performed, with a single hearing. If need be Sarah could write the notes down half a century later. Mozart had the gift, but few others do. She sometimes entertained me by singing long ago pop tunes and jingles from long forgotten 1950's television commercials.

## MUSICAL GIFTS

In Spokane she was the bishop's cantor, which made her a celebrity in that very Catholic town. It became difficult for her to go out in public without being importuned. Eventually, the constant attention and lack of privacy drove her to Southern California where she could hide in its vast crowds. Fortunate for me, of course, otherwise we wouldn't have met.

# THE RAID

EACH YEAR, AT CHRISTMAS AND EASTER, Sarah teamed up with a couple of singing FBI agents. The three of them toured the local prisons to cheer up the inmates.

One day Sarah was reading when a disturbance outside caught her attention. Skulking through the yard were men, guns drawn, wearing FBI jackets. They were surrounding the house next door. It was a raid. Sarah knew what was happening for she had told her FBI friends about the place.

We didn't know these neighbors. We had never seen them for they were total recluses. All we knew when we moved in, a year before, was that the house was rented.

A nearby neighbor was a bit of a sleuth. She had figured out what was going on with these never seen people. It was the mysterious irregular traffic on our out-of-the-way street that raised her suspicions. So she kept careful watch. The unusual traffic and the unsavory appearance of the visitors informed her. The place was a crack house. The people there were peddling hard drugs. The signal was the garage door. If it was down, stay away. If it was up the coast was clear and they were open for business.

After Sarah alerted the FBI the nest of snakes was cleaned out. The owner of the house, living back east, had not known about these renters. After this incident he

## THE RAID

was much more careful who he rented to so we had good neighbors from then on. Eventually the house was sold and things settled down for the long haul.

# THE HORSE

**WHAT IS IT WITH A WOMAN AND HER HORSE?** I had run into the phenomenon before and Sarah still had it bad, if only in memory. When she was young she owned a horse named Dick. It had a black coat with a white star on its forehead. Dick stayed at her grandmother's house down in Georgia. But it was Sarah's when she visited every summer.

Her family had been burned out when Sherman marched through Georgia. Everything was lost except the land. They slowly rebuilt but Sarah remembered that when she was very young the small house still had a packed down dirt floor. And some of the animals were brought inside during the winter to provide extra heat. It was a big day when her grandparents had prospered enough to put wood flooring in and install adequate heating. Sarah remembered that, too.

As a child, Sarah had some problems with her teeth. The solution was orthodontia. Her orthodontist was Dr. Z, as we called him. Sarah being the charmer that she was the two instantly became buddies. Long after the problem with her teeth had been fixed Sarah and Dr. Z stayed in close touch.

Now Dr. Z had a ranch some distance out in the country. One day we paid him a visit. The place was special. Dr. Z, in addition to being a very good orthodontist, was also the Mormon bishop for the region. Following in the tradition of Mormons, his place was immaculate. In fact his stable was washed so clean I had the impression you could do brain surgery in it without harm to the patient.

Dr. Z raised horses. Arabians. Beautiful animals. In one corral a pair of gray Arabians were overseeing the antics of their recently born colt. That youngster was frisking around the corral full of the joy of new life. Sarah and I instantly fell in love with the bundle of energy.

Dr. Z, observing our reaction, was amused. What he said then was unexpected. He offered the young stallion to us as a gift. No joke. He was sincere in this most generous offering. I knew there was no way we could possibly take such a valuable animal. Sarah understood that too. But Sarah, being Sarah, had to play the role. She started begging me to take the colt. It was part of her comic act, of course, and Dr. Z, on the sidelines, and knowing Sarah, realized this. He could no longer keep a straight face.

"Please," Sarah begged with a bit of a whine. I scolded her. We had no place to keep a horse. "We could drain the swimming pool and keep it there," she pleaded. The three of us ended up laughing. The colt stayed in the corral.

# GENERAL RICHARDS

Sarah and I honeymooned in Yosemite. On the way home we visited the Giant Sequoias in Kings Canyon National Park. We had brought Sarah's orange tabby, Sunny, along with us on our honeymoon. Sunny slipped out of the car when I opened the door and ran off into the woods, making a bee line for one of the largest sequoias.

What had caught her attention was the large cavern in the center of the tree. Some old-time burn had hollowed out the base of the tree but otherwise did little harm. The great tree continued its long life as if nothing much had happened. The tree was behind a protective fence but with Sunny in that cavern, refusing to come back to my call – most unlike that cat – I had no choice but to climb over the fence and enter the cavern myself.

Inside that great tree was a good-sized room with a full blown campsite, festooned with sleeping bags, a camp stove and miscellaneous gear. Sunny, her curiosity satisfied, was happy to accompany me back to the car. I left her with the windows cracked open for ventilation. There was no chance of overheating in the cool mountain air. Sarah and I wandered off into the woods to commune with the great giants.

When we got home and were unloading the car I discovered that Sarah had pilfered some sequoia seed cones — a definite no-no. She had done something like this before when she smuggled a fuchsia seedling back from Seattle (it's still growing, thirty years later). California has strict rules against such uncontrolled imports. Sarah had little regard for excessive restrictions. Seed cones of rare trees should properly be left in place, though. Who knows what the future would hold for them? They might become the future's giant trees. But what was I to do? I wasn't about to drive the hundreds of miles back to drop off a couple of seed cones. I let the matter go.

A couple of years later Sarah decided to experiment. She read somewhere that sequoia seeds required fire to germinate. Apparently they must be scorched but not completely burned up. She put one of them in the fireplace to see what would happen. Sure enough the outer portion charred leaving the core intact. This she planted in a suitable spot in the backyard. In time a seedling appeared and began to grow, rapidly. Following the tradition of naming the giant sequoias after generals — General Lee, General Grant, etc. — we dubbed our youngster "General Richards." General Richards became a fine looking stripling.

Years passed and we added Hector to the family. As you know (see Cats and a Dog) Hector grew into a fine German Shepherd. Very headstrong, but loyal and very smart. One day I looked out the window. There was Hector, happily prancing around the yard with General Richards in his mouth, the seven foot tall tree, roots and all, projecting sideways for several feet in both directions. So much for the general in our family — at least for some years. My sister later had that rank, O-8, but that is a very different story.

# SCARY

We left our lodging for the night well before dawn and grabbed breakfast at a coffee shop along the way. The place was open early. It was autumn and harvest season so working people were headed out to the fields early to make best use of the shorter days. Sarah and I headed for the high country.

We had reached a mountain meadow when the sun had just begun poking up beyond the far rim. Suddenly we were in a wonderland. The fields around us were gleaming, aglow from a carpet of sun-sparkling diamonds. A heavy frost had settled during the night and coated the long grass with rime. We stopped. We just had to take photos of this.

I was down on my belly shooting a close-up when I involuntarily froze. Echoing around the distant low hills was a sound that cut through me to the marrow in my bones: The sound of an agonizing howl interlaced with the screech of metal being tortured. What was that? Sarah was nonchalant. "Oh just a timber wolf cutting loose," she replied. "It's hungry and it's time to find some breakfast."

Sarah had heard the sound many times before. Long ago when she was a National Park Service ranger, living alone in a fire watchtower on top of a mountain in northern Idaho. Our photos taken, we got back in the car and drove on.

# WINGNUTS

SARAH AND I HAD SETTLED in a town that serves Hollywood. Mostly the Hollywood people work behind the scenes: cinematographers, artists, set builders and decorators, technicians of various kinds and people who move things from here to there. I know a studio casting director who lives nearby, and a house down the street has appeared in several movies. Actors live here, too. Every once in a while I will see a familiar face from the big screen, or the little one, in the supermarket or the bookstore. For more than a year a noted character actor rented the house next door. He was appearing in a popular TV soap opera that was in production nearby.

Then there are the stuntmen. Forget the fame of celebrities, the real royalty in Hollywood are the stuntmen. Sarah sang at the funeral of one of them, killed during filming. She was the cantor for the mass. The crowd in attendance was enormous. All of Hollywood turned out. TV screens had to be set up outside the church for the overflow filling the streets.

If Hollywood people live here so do their kids. Their kids go to school. And they talk to each other. We were to learn how fast an idea can take hold and propagate from the schoolyard to Hollywood and thence to the country and then the world at large.

In the state of Washington Sarah was a licensed psychotherapist. But California does not have reciprocity with other states. So Sarah had to go through the California Marriage, Family and Child Counseling licensing process. This gave her both the MFCC license and another Master's degree. (At least three, one new one in Psychotherapy, to add to her old ones in Music and English Literature, and also maybe one in Psychology in the State of Washington.)

Now having her MFCC, Sarah counseled in a local high school. This school served that part of town that had a high concentration of Hollywood people. Some of those kids really need counseling. Hollywood is a very high pressure industry and it takes its toll on the families.

One evening Sarah and I were watching some oddball making a fool of himself on TV. Sarah called him a "wingnut." Of course I knew about this handy type of fastener but I had never heard the term applied to a person before. "Where did that come from?" I asked. "Oh, it just seemed the appropriate thing to say." Sarah had just made it up. Over the next couple of days we found lots of opportunities to apply the term to manifestly silly people on the air.

I suggested an experiment. Why not use the term during her next group counseling session at the high school. She did and it caught on and spread like wildfire. The term was just so appropriate to describe a whole bunch of people. A couple of weeks later we were watching a national television show when the presenter called someone they were talking about a wingnut. The kids had taken the expression to their parents and they, in turn, took it to Hollywood. Hollywood gave it to television and it spread across the nation from there. The appellation went from our darkened TV room to spanning the nation in two weeks flat.

Shortly thereafter *Newsweek Magazine* published its annual lists of new words that had been added during the year. Wingnut was at the head of the list. Then the "Lord of the Rings" movie trilogy came out of New Zealand. The production studio Down Under was named Wingnuts. Sarah's little inspiration had spanned the globe and now you could find it in all English dictionaries. Eventually the fad for the word diminished and *wingnut* became just one more word in the vocabularies of seventy languages, world-wide.

I HAVE REMAINED IN THE HOUSE Sarah and I once shared. Occasionally, I meet or read about a new wingnut. Sarah chuckles softly from the wings.

*From our garden*

# ABOUT THE AUTHOR

**CHESTER L. RICHARDS,** retired aerospace engineer, and inventor — 19 patents, #20 pending — views life as a series of adventures. In surfing, learning and performing music, travel to exotic places, white water rafting, work as a rocket scientist. And writing — he's penned over 60 essays, 10 more since his first book.

Along the way, he's always met fascinating characters. But it was the loss of Sarah, the love of his life, that caused Chester to write his first book of stories, the award-winning *From the Potato to Star Trek and Beyond: Memoirs of a Rocket Scientist.* In this second volume in the planned trilogy, titled *The Trek Continues,* you'll find more stories of the hair-raising adventures and groundbreaking projects Sarah — and now you, Dear Readers — loved, and more about the amazing lady herself. You'll meet more extraordinary people, plus phenomenal cats and a miracle dog named Hector. The author thanks you for your kind response to "The Potato," and hopes you enjoy traveling with him to new narrow escapes from the jaws of death, and musings on his life's theme: every day is an adventure.

# PHOTO CREDITS

Photo of Star Trek credits scene "Written by Judy Burns and Chet Richards," page xii. Screenshot from show.

Whitewater rafting — Big Hole at Crystal, Colorado River, Grand Canyon, page 10. Photo by Author. © 2025 Chester L. Richards

Mursi Tribesmen still Neolithic, Omo Expedition 1978, page 12. Photo by Author. © Chester L. Richards

A Nile crocodile on the bank of the Omo, page 17. Photo by Author. © 2025 Chester L. Richards

Runup to Omo Expedition, Awash River, page 18. Photo by Author. © Chester L. Richards

Monkey on monkey time among Vervets, page 22. 20397845 © Na8011seein. | Dreamstime.com

Who could resist, page 23. Photo by Author. © 2022 Chester L. Richards

Vultures in the sleeping tree, page 24. Photo by Author. © 2022 Chester L. Richards

Gelada Baboon. Awash, Ethiopia, roaring his alarm, page 31. 50637323 | © Zlikovec | Dreamstime.com

PHOTO CREDITS

Hot springs oasis in the midst of the desert, page 33. Photo by Author. © 2025 Chester L. Richards

Family of warthogs, page 36. Photo by Author. © 2022 Chester L. Richards

A Flying Beetle, page 58. Photo by Author. © 2025 Chester L. Richards

Avalon Harbor, Catalina Island, page 74. 4454941 © Rraheb | Dreamstime.com

A herd of Buffalo innocently grazing on Catalina Island, page 103. 652279. © Heather Jones | Dreamstime.com

Chester at sunset after a day of surfing at the beach, page 104. Photo by Author. © 2025 Chester L. Richards

Fireworks photo collage, page 116. Vintage lady finger firecracker. https://i.pinimg.com/originals/1f/6a/3a/1f6a3a2b6d050ce99a8ff86c817a6927.jpg. Snake burning. https://i.ytimg.com/vi/OisFvFpv15I/maxresdefault.jpg

Chester getting treated after scorpion bite, page 119. © 2025 Chester L. Richards

A U.S. Air Force Fairchild C-119B-10-FA Flying Boxcar of the 314th Troop Carrier Group in 1952, page 120. (U.S. Air Force photo).Public Domain, http://commons.wikimedia.org/

Peter the Anteater™, page 142. © Regents of the University of California

THE TREK CONTINUES

Just like Chester's mid-70's VW, page 193. VW © Marcin Melkis | Dreamstime.com. Beach background photo © 1957, 2025 Chester L. Richards

Woolsey Fire as smoke billows over the top of the mountain, page 206. Photo by Author. © 2025 Chester L. Richards

Sparks reading about System Architecture, page 214. Photo by Author. © 2025 Chester L. Richards

Hycon Patents, pages 219 & 220.
Original images in patents and photos on pages © 2025 Chester L. Richards.
Automatic Focus Sensor. https://patentimages.storage.googleapis.com/a8/67/ec/bd2a9d301e2d8d/US3555280.pdf
Pneumatically Operated Camera Shutter. https://patentimages.storage.googleapis.com/7d/92/33/ff0d1ac3f14ae6/US3502015.pdf

Space shuttle Atlantis is seen as it launches from pad 39A on Friday, July 8, 2011, at NASA's Kennedy Space Center in Cape Canaveral, Fla, page 274. Photo Credit: (NASA/Bill Ingalls). https://publicintelligence.net/wp-content/uploads/2011/07/sts-135-16.jpg

Close-up side view of Space Shuttle Main Engine, page 281. Courtesy of the Library of Congress. https://www.loc.gov/pictures/item/tx1115.photos.579959p

Mom and Dad, page 284. Family photo collection. © 2022 Chester L. Richards

Grandmother Richards (Alta Sayles Richards), page 286. Family photo collection. © 2025 Chester L. Richards

PHOTO CREDITS

Great-Grandfather Truman S. Richards, page 291. From family archives. © 2025 Chester L. Richards

Christmas Day 1942, page 298. From our family photo collection. © 2025 Chester L. Richards

My sister Joan Marie, page 299. From our family photo collection. © 2025 Chester L. Richards

Mom in her kitchen, page 300. From our family photo collection. © 2025 Chester L. Richards

Apple pie, page 307. Photo by biglanphoto via iStockphoto.com

My dear Sarah, page 308. Photo by Author. © 2025 Chester L. Richards

The love of my life Sarah, and me, at our wedding, page 310. Photo © 2025 Chester L. Richards

Sarah, my amazing bride, and my wonderful sister Marie, at our wedding, page 314. Photo © 2025 Chester L. Richards

Author Chester L. Richards holding his and his wife Sarah's amazing dog Hector, page 318. Photo by Author. © 2025 Chester L. Richards

Merlin and Misty in cat bed, page 330. Photo by Author. © 2025 Chester L. Richards

Sparks, page 341. Photo by Author. © 2025 Chester L. Richards

THE TREK CONTINUES

Merlin bathing Sparks, page 343. Photo by Author. © 2025 Chester L. Richards

Sarah with our cat Sunny, page 344. Photo by Author. © 2025 Chester L. Richards

From the garden at our house, page 358. Photo by Author. © 2025 Chester L. Richards

The author, Chester L. Richards, page 359 Photo by Chris Schmitt Photography, https://www.chrisschmitt.com. © 2025 Chester L. Richards

www.ingramcontent.com/pod-product-compliance
Lightning Source LLC
Chambersburg PA
CBHW070734170426
43200CB00007B/519